Patricia McCarthy Veach
Bonnie S. LeRoy
Dianne M. Bartels

Facilitating the Genetic Counseling Process

A Practice Manual

Springer

Patricia McCarthy Veach
Department of Educational
  Psychology
139 Burton Hall
University of Minnesota
Minneapolis, MN 55455, USA
veach001@umn.edu

Bonnie S. LeRoy
Graduate Program in Genetic
  Counseling
University of Minnesota
420 Delaware Street, SE, MMC 485
Minneapolis, MN 55455, USA
leroy001@umn.edu

Dianne M. Bartels
Center for Bioethics
University of Minnesota
Suite 504 Boynton
410 Church Street, SE
Minneapolis, MN 55455, USA
barte001@umn.edu

Library of Congress Cataloging-in-Publication Date
McCarthy Veach, Patricia.
    Facilitating the genetic counseling process : a practice manual / Patricia McCarthy Veach
Bonnie S. Leroy, Diane M. Bartels.
      p. cm.
    Includes bibliographical references and index
    ISBN 0-387-00330-4. (softcover : alk. paper)
    1. Genetic Counseling—Handbooks, manuals, etc. 1. Veach, Patricia McCarthy. II.
Bartels, Dianne M. III. Title.

RB155.7.L47 2003
362.1'96042—dc21   2003042477

ISBN 0-387-00330-4   Printed on acid-free paper.

Printed in the United States of America

9 8 7 6 5 4 3 2 1   SPIN 10906767

www.springer-ny.com

Springer-Verlag New York Berlin Heidelberg
A member of BertelsmannSpringer Science+Business Media GmbH

Preface

Overview

This manual is intended to help genetic counseling students develop basic helping skills that form the foundation of effective genetic counseling relationships. The chapters address several of the psychosocial, practice-based competencies established by the American Board of Genetic Counseling (1996) (see Appendix A for a list of the ABGC Clinical Competencies), including: Communication Skills (Chapters 2-7), Interpersonal Counseling and Psychosocial Assessment Skills (Chapters 8-10), and Professional Ethics and Values (Chapters 11 and 12). The content contained in this manual will not fully prepare students to practice independently. However, it will provide a skill base that deepens and broadens as students gain additional academic and clinical preparation. Although we designed this manual for use in a classroom setting, most of the materials, activities, and exercises can be adapted by clinical supervisors for use in the clinical setting.

We drew upon a number of sources in writing this manual, including our combined professional experience as practitioners, educators, and researchers in the fields of mental health counseling, genetic counseling, psychosocial nursing, family social science, and bioethics, and from literature in the fields of psychology and genetic counseling. Many of the articles and books that we cite are from the mental health counseling/psychotherapy field. There are striking similarities between mental health counseling and genetic counseling, and many concepts are virtually interchangeable. However, our goal is not to train genetic counseling students to be psychotherapists. Therefore, the examples that we provide to illustrate skills, concepts, and processes are always specific to the genetic counseling relationship.

The content is grounded in a variety of theoretical presentations, including: humanistic theories that stress the significance of helper genuineness, positive regard, respect, and nondirectiveness; psychodynamic theories that emphasize the strength and quality of the helper-client working alliance and conscious and unconscious processes; and cognitive-behavioral theories that describe the complex interaction of thoughts, feelings, and behaviors and the importance of defining client concerns and goals in concrete, behavioral terms.

Our perspective is strongly influenced by our white, Western cultural backgrounds and by the tenets of traditional Western medicine. We attempt to broaden this perspective by pointing out the limitations of certain concepts and techniques for clients whose cultural practices, beliefs, and worldviews differ from our own. In addition, we include some examples of clients with diverse backgrounds. It is important to keep in mind, however, that a few examples do not represent all cultural groups, nor do they apply to all of the members within a particular group.

Manual Format and Philosophy

We begin each chapter by stating general learning objectives. Then we define the skills, place them in a context (their function or purpose in genetic counseling), give examples, and provide activities and written exercises for skill practice. We conclude each chapter with a brief annotated bibliography to provide additional resources on major topics. The reader may notice some redundancy in the examples (e.g., many involve prenatal genetic counseling situations; and Down's syndrome, muscular dystrophy, and Huntington's disease are frequently mentioned). This redundancy is intentional because we wanted our examples to be "basic" enough that students at all levels, including those with limited knowledge about genetic conditions, would be able to understand the examples and participate in related activities.

Because we believe that guided practice and critical reflection lead to skill improvement and to increased self-awareness of significant professional issues, we provide a number of written exercises and structured activities. The exercises and activities are based on an active and cooperative learning approach that emphasizes a high level of student participation and student responsibility for learning (Johnson, Johnson, & Smith, 1991). Students work together in learning activities that promote cooperation rather than competition. Furthermore, this approach facilitates student self-assessment of strengths and limitations. A great deal of research supports the superior outcomes that result from active, cooperative learning (Johnson et al., 1991). Moreover, reading and/or lectures alone are not sufficient for developing effective genetic counseling helping skills.

Most of the written exercises can be modified to use as structured activities in the classroom or clinical setting. The structured activities can be done before the written exercises to stimulate student thinking, or after the written exercises to afford students the opportunity to think through how much they are comfortable learning. Regardless of which exercises and activities are chosen, and whether they are done in writing or orally, **students need to be cautioned to select only those issues they are comfortable disclosing to others.** Instructors/supervisors should reinforce this point and always inform students in advance about the types of information they will be expected to share with others.

Closing Comments

We ask that you begin with the chapter, "Guidelines for Manual Users," as it "sets the stage" for the remaining chapters. We hope that you find this manual useful, and we welcome any comments or questions you might have about it. We can be contacted at: veach001@umn.edu; leroy001@umn.edu; and barte001@umn.edu.

Patricia McCarthy Veach
Bonnie S. LeRoy
Dianne M. Bartels

Acknowledgments

Preparation of this manual was supported by a grant from the Jane Engelberg Memorial Fellowship.

We were fortunate to have the input of the following individuals who served as advisory board members to our project: Diane Baker, Elisabeth Gettig, Caroline Lieber, Janice Edwards, Bonnie Hatten, Catherine Wicklund, Jane Schuette, Nicole Teed, Rob Finch, Annette Patterson, Mary Ann Whalen, Joyce Weinsheimer, and Vivian Ota-Wang. These individuals contributed their expertise from the perspectives of the genetic counseling program director, clinical supervisor, advanced genetic counseling student, education specialist, and cross-cultural counselor. They provided critical comments on chapter drafts and shared a number of excellent, practice-based examples that we included in the manual. We would also like to acknowledge the assistance of the following individuals in the development of this manual: Matthew Bower, Matthew Hanson, Rafael Robert, and Pam Folkens.

Contents

Contents

CHAPTER 1

Guidelines for Manual Users: Instructors, Supervisors, and Students

Helping skills are fun to teach and to learn. Students are eager to learn the techniques of their craft, and they enjoy the variety of activities involved in skills practice. At the same time helping skills training involves some unique challenges (e.g., distinguishing between similar types of counselor responses such as primary empathy and advanced empathy, reducing student anxiety and resistance, and differentiating developmental issues from issues of skill deficiency). This chapter offers suggestions for teaching and learning helping skills and discusses some of the challenges involved in helping skills development. We describe several active learning techniques that are appropriate for helping skills training and offer tips for feedback, facilitating discussion, and conducting role-plays. We give suggestions about evaluation and student resistance, and we include three student activities for integrating learning (see Appendices 1.3, 1.4, and 1.5). Most of these suggestions can be adapted by clinical supervisors to use with their supervisees. Many of these ideas have been shaped by the students and colleagues with whom we have worked over the last 20+ years. We are particularly grateful to the Center for Teaching and Learning Services at the University of Minnesota.

Active Learning Guidelines and Techniques

As we discussed in the Introduction, this manual is grounded in an active and cooperative learning approach to skill mastery (Johnson et al, 1991). In this section we begin by making general suggestions for using an active learning approach. Then we provide examples of different types of active learning techniques.

General Suggestions

Get Started

- Describe the active learning philosophy and how it relates to your learning objectives. We include a description of active learning on our syllabus and discuss it during the first class period. For some students, this may be the first time they have participated in a course that it is not primarily in the lecture format.

- Begin the first class with an "ice breaker" active learning exercise. This sets the tone for the types of activities that will occur throughout the course. For example, in the "note cards" ice breaker exercise, students write down on index cards personal information such as their name, home town, favorite book or movie, and one or two things they hope to learn from the course. Then they walk around and discuss their information with others in the class.

Build the Relationship

- You and your students should learn each other's names as quickly as possible if you don't already know them. For instance, use name tags; or play a "name game" in which you go around the circle and each person says her or his first name and a self-descriptive adjective beginning with the first letter of her or his name (e.g., athletic Annie), the next person says her or his name and adjective and repeats the name and adjective of the previous person. Continue this way around the circle until the last person (perhaps the instructor) has to repeat everyone's name and adjective.

  Vary the way students join dyads, triads, and small groups so that they have the opportunity to interact with everyone (e.g., count off; preassign; everyone who is at the same table; let students pick a partner—especially appropriate early in the course and/or for activities where students might disclose highly intimate information).

Stay Focused

- Provide verbal and written directions for all activities (provide handouts, put directions on an overhead transparency, and/or write them on the board).
- Ask a student to verbally summarize your directions for an exercise.
- Earlier in the course you will need to provide more structure and instructions than you will later. For instance, groups may not naturally engage in necessary activities such as keeping time, recording group member ideas, and working to include everyone in the conversation. You should assign essential roles for small group discussions (e.g., go around the small group and say, "The person whose last name is the shortest will be the recorder, the person to her or his left is the time keeper, the next person is the process observer, the next person is the divergent thinker, the next person is the facilitator, the next person is the reporter").
- Walk around during active learning exercises to get a feel for what is developing, to help keep students on task, and to clarify instructions. Inform students that you will be "listening in" throughout the course; they will quickly acclimate to having you walk around or sit in with them.

- Move people along, especially since individuals would rather talk than practice. For instance, say, "I know there is more we could discuss, but I want to be sure that you all get a chance to practice the skills, so let's take one more comment before we move on."

Be Efficient

- For small group activities vary the way roles are assigned so that they are determined quickly (e.g., the recorder is the person wearing red, or the person with a birthday closest to that day, or the tallest person, etc.).
- When debriefing an activity in which more than one small group discussed the same questions, to avoid redundancy have each group give one idea, or have each group give its answers to a different part of the question. Keep going around until all ideas have been expressed.
- When planning activities, be sure to allot time for instructions and for students to get into work groups. We provide time estimates at the end of each chapter. However, the time for activities will vary considerably depending on (1) the size of the class, (2) the verbosity of the students, (3) the number and quality of questions you use to process an activity, and (4) the complexity of the skill or concept on which the activity is based.

Tips for Instructors

The activities in this book emphasize self-reflection, discussion, and skills practice. To maximize these experiences, we suggest the following:

Facilitating Discussion

Responding to Student Questions and Answers

- Occasionally redirect questions from students back to the group, but only if you believe that someone will have a good answer.
- Be selective in what you reinforce. Try to relate everyone's answers or comments to the issues at hand. Repeat the most pertinent or useful comments in a summary statement.

Encouraging Student Participation

- Watch for nonverbal behaviors you can use to draw individuals into the discussion.
- Be sensitive to individual differences. As you get to know your students, you will want to vary how you bring them into discussions. For example, if a student never volunteers and seems reluctant, occasionally invite this student to give a reaction first during a discussion. Or, if a student is verbose, ask for that student's feedback last.

USING SMALL GROUPS

- When using a small group format, four to five students per group are an optimal number for encouraging participation and generating quality discussion.

- To facilitate discussion, start with questions that anyone could answer, then make them progressively more difficult.

Example: You can begin a discussion about "relationships" by asking everyone to respond to the question, "What are they?" Then ask more specifically about what the genetic counseling relationship entails, its positive and negative aspects, and the goals of the relationship.

- To maximize small group discussion, first define the concepts or terms that will be discussed, and provide a brief overview of the material. In processing the discussion, try to *tie together* student comments by summarizing major themes and issues. Also be prepared to correct any inaccurate information that may emerge.

Promoting Skills Development

USE EXAMPLES

- Provide as many examples as you can when presenting material. Novices are extremely interested in seeing how things are done. One technique you might use is to refer students to places in the text where we provide an example and ask them to generate several more. This will facilitate their learning and comprehension.

- Concrete examples are very helpful for illustrating concepts. If you can, provide students with videotaped, audiotaped, and live demonstrations of genetic counseling (preferably by more than one genetic counselor). If feasible, bring in volunteers to serve as genetic counselors and clients for some of the demonstrations.

- Make your examples basic enough that students do not need a lot of knowledge about the genetic condition. Provide them with some details about the condition so that they are able to proceed with the activities.

ORGANIZE CLASS SESSIONS

- When preparing each class, prioritize activities so that you know in advance which ones you will delete if you run overtime.

- Arrange your class activities so that they progress from easier to more challenging ones. You should also begin with less threatening activities (e.g., defining *defense mechanisms*) and then move to more threatening activities (e.g., discussing one's own defense mechanisms). When arranging activities, remember that the more threatening an activity, the fewer people you may want to have listening to a student's disclosure (e.g., use a dyad where students select who they want as a partner). When processing an activity, don't ask for details, although students are free to offer them. For instance, in processing a defense mechanisms dyadic exercise, ask, "How was it to do this activity? What did you learn about the impact of defense mechanisms on genetic counseling?" Do not ask, "What defense mechanisms do you use?"

- Prepare and have on hand several "generic" role-play scenarios that you could assign to students for role-play practice. You can also use the scenarios from the exercises and activities at the end of the chapters.

- If feasible, use co-instructors (e.g., advanced students). They will provide different viewpoints, and you may have enough co-instructors to directly observe all student role-plays. Ideally there would be one instructor for each triad and small group activity. Co-instructors can also serve as counselors and clients when demonstrating helping skills.

DEMONSTRATE/MODEL

- One way that individuals learn is by contrast. When time allows, model both low-level and high-level helping skills, always beginning with low levels. Ask students to articulate the differences between the two levels.

- Use processing questions after a demonstration. For instance: "What did you observe? What effect did it have on the client? What does it mean? Is it desirable? Undesirable? What would you have done differently and why?"

- You can set norms by going first and modeling how to do an activity.

Strategies for Facilitating Role-Plays

A Role-Play Model

Role-playing is the primary learning activity for the skills covered in this manual. There is no single way to conduct and process/debrief role-plays (see Appendix 1.1 for a description of role-playing). In addition, we recommend the following:

1. Organize students into triads where they take turns being the counselor, client, and observer (change triad composition frequently).

2. Remind students of how much time they have for each role-play.

3. Ask for volunteers to go first as the counselor and client.

4. Remind observers to take notes and to keep track of time.

5. Have the counselor and client position themselves as if it were an actual genetic counseling session (they may have to move chairs).

6. Direct student counselors to focus on all of the skills they have covered so far and use them *as appropriate* (in other words, don't force a skill just for the sake of demonstrating it).

7. Tell counselors that they can call for a *time-out* during the role-play if they get really stuck. The observer can also call a time-out if things seem to be bogged down. During the time-out the counselor should talk about what s/he thinks is going on (what the client has been saying, doing, feeling), and the counselor and observer can consult about ways for the counselor to proceed. The client should be silent during the time-out. Then resume the role-play (it usually helps to have the client begin). When there is a time-out, reduce the amount of feedback time at the end of that role-play.

8. Debrief by having the observer give at least one positive and one corrective piece of feedback. Next ask the client to provide feedback. As the group gains experience during the course, debriefing can begin with the counselor providing a self-critique and then proceeding to observer and client feedback.

9. Remind students that feedback should *focus on the counselor and not the client!*

10. Remind students to focus their feedback on the skill for that session and any skills that have been covered prior to that session. Try to minimize feedback on skills that have not been covered (This is especially likely to happen in early class sessions; for instance, students are practicing attending skills but will give feedback about questioning skills).

11. Some clients get caught up in role-plays and may become emotional. Let them regain composure before eliciting their feedback. Also, depersonalize feedback to the counselor that involves the client since some elements of the role-play are likely to be the student/client's real reactions and/or history. For instance, say, "When clients are highly defensive, it's a good idea to..." Do not say, "Joan was a highly defensive client, so you should have..." Or say, "Your use of open questions with this type of verbal client was..."

12. Sit in and observe each student during role-plays as much as possible during the course.

13. Once students have participated in a few role-plays and have a sense of their current skill level, you can invite them to request feedback on certain skills in advance of the role-play. Feedback is most effective when it is requested.

14. If feasible, videotape students during some role-plays. They will likely feel anxious about this, but will learn a great deal from seeing and hearing themselves.

15. Be sure that everyone's role-play observation notes are shredded and discarded at the end of each class.

CRITICAL ISSUES IN ROLE-PLAYING AND DEBRIEFING

• As we stated earlier, people would rather *talk* than *do*. You can easily get off-schedule, talking about the skills and not having enough time to practice. Encourage students to practice.

- The counselor and/or client get off track during the role-play. When this happens, the observer should call for a time-out.

- Time is running out. If you wish to limit discussion, have the observer and client give only one or two pieces of feedback to the counselor. The role-plays could also be shortened by a couple of minutes.

- Students provide invalid and/or harsh feedback. Sit in on role-plays and model for students how to give feedback. If you openly disagree with a student's feedback while sitting in on a role-play, be tactful (e.g., "I think I had a different reaction to the counselor's approach to this client. I think this shows how different clients might react differently to the same counselor behavior"). Another option is to ask the other student in the triad (either the client or the observer) if s/he had a similar reaction to that of the feedback giver.

- The counselor is defensive. Remember to use the basic helping skills—a little empathy goes a long way! Also, put feedback into a *context* for the student (e.g., "This is something most beginners do," or "This isn't a big deal," or "With practice, you'll improve on that behavior"). Role-playing is a threatening activity, so expect some anxiety. In our course evaluations, students will tell us it's the activity they dreaded the most, but that it is also one of the activities they got the most out of in the course (They respond similarly to the self-critiqued, audiotaped role-play assignments). Also, the most experienced students are often the most nervous about role-playing. Perhaps they believe that more is expected of them.

- Discrepant feedback. Students will likely hear contradictory feedback from different observers, and they may become frustrated or confused by this. We tell our students that they should listen for the *themes* in the feedback they receive. One isolated comment that they were too directive may not be as valid as several comments from different sources. Contradictory feedback may be particularly troublesome for some novices who are looking for formulas or the *right way* to do things.

- Students complain about using made-up material during role-plays. Some students complain about the artificiality of simulated role-plays (e.g., it's not how it would really happen; they couldn't really *get into it* because they knew it wasn't real). We acknowledge that there is a certain degree of artificiality. We also talk about how practice is important (e.g., student nurses administer shots before they see actual patients) and encourage students to try for as much realism as possible. Furthermore, we believe that once students get over some of their initial anxiety about being observed, they settle into role-playing. We point out that it is very difficult to construct a completely hypothetical role. The role-player will project her or his own feelings, thoughts, and attitudes into the role.

PROVIDING FEEDBACK

- Provide a balance of positive and corrective feedback. It can help to begin with the positive. Always suggest what the student might try in order to improve. Try the *sandwich technique:* tell the student what s/he did well, next suggest areas to work on, and finish with a reiteration of what s/he did well.

- Ask students to self-evaluate.

- When students give each other feedback, tell them to talk directly to the person receiving the feedback and not to the instructor.

- When giving feedback, students may go to extremes—only talking about the positive aspects of another student's role-play (e.g., "You did everything just great!"), or hammering another student with a laundry list of everything the student did wrong. We recommend that you discuss giving and receiving feedback at the beginning of the course and use feedback exercises (described in Appendix 1.2).

Additionally, Flash et al (1995) offer the following feedback suggestions:

- Provide feedback frequently.

- Give clear, specific feedback that offers guidance on how to improve.

- Consider possible cultural differences (e.g., when evaluating writing by students for whom English is a second language).

- Tie your feedback to your course objectives.

- Give students an opportunity to use your feedback to improve their performance (e.g., allow them to turn in drafts before a final written product is due).

Selected Techniques

Active Learning Exercises

The following list contains a sampling of different types of active learning exercises that might be appropriate for your setting and learning objectives:

- Survey the class: "How many of you agree with the author's point of view? How many disagree?" Have students raise their hands.

- Random calling: For larger classes, randomly call on individual students or dyads (e.g., write the name of each student on Popsicle sticks. Randomly draw a stick from a container and call on that student).

- Bean counters: In small groups, everyone gets three beans or three poker chips, and each time a person speaks, s/he throws a bean into a bowl or box. When a person's beans are gone, then s/he can no longer speak.

- Speaking stick: Based on Native-American practice, the stick is passed among the group members. Whoever has the stick is the only one allowed to speak.

- Margin it: Students write down answers or responses in their notebook margins. This is a safe, anonymous way to check themselves out on what they know. After doing this, you provide the answer or have volunteers share what they wrote.

- Think-pair-share: This is a dyadic activity. Students first think about a question or concept. Next they find a partner, and discuss responses with each other. To process, you could go around and ask each dyad to discuss one idea until the concept or question has been fully explored. One variation is to have students write down their response before talking with a partner (e.g., "Write down everything that you know about empathy"). Another variation is to have one dyad join another dyad. The resulting quad discusses responses to the questions.

- In-class writing: Give students 1 to 5 minutes to "Take a stance," "Defend a position," or "Formulate a response to the following client statement..." Then have students discuss what they wrote with a dyad partner, small group, or the whole class.

- Data interpretation: Have dyads or small groups read and interpret graphs, tables, or charts. For example, they could read a table of data about the risk for a particular genetic condition, interpret the data, and formulate a way to explain this risk to a genetic counseling client.

- Laundry list: Students raise all of the questions they have about a topic (e.g., touching clients, nondirectiveness, self-disclosing with clients) and you list them on the board. Then you proceed to address all of the questions that are raised (using whatever format is appropriate—lecture, in-class writing, dyads, etc.).

- Ten- to 20-minute press conference: Students write down anonymous questions about material or issues covered in the course so far. You collect them, shuffle them, and redistribute them. Then students read any interesting questions, and you attempt to answer them. We have used this format at the end of a course. By then students are quite comfortable raising complex issues and sensitive topics. However, it should also work well earlier in a course. This exercise is particularly effective if your course format includes co-instructors—students can hear different opinions to the same question. One caveat: tell students that the questions should be about course content and not format. (We discourage questions such as, "Why is assignment 1 worth so many points?" This type of question is more appropriate for course/instructor evaluations.)

- Student originated questions/cases: Assign students the task of finding challenging genetic counseling situations in the literature and then bringing them to class to work on in groups (e.g., strategizing how to respond, role-playing the situations, discussing them).

- Dialogue journals: Students pair off and respond to each other's journal entries for 5 minutes of each class. These dyads stay intact for the whole course. You can assign them journal topics to discuss or leave it fairly open-ended. Require that journal entries be focused on the course. You should periodically collect the journals and informally look at them. It is also a good idea to provide examples of journal entries so that students have some idea regarding length, scope, and topics.

- Matching and milling: Individual students are given a piece of information and then they move around the room comparing their information with the other students' information. Their task is to figure out how the pieces fit together. For instance, you could give each student pieces of a client's family history, including some relevant and nonrelevant information. The students' task is to figure out the client's risk. Next, groups of four to five students could brainstorm and then role-play how to convey this risk information to the client.

Example: Construct a complicated family history for cancer. Assign some class members to portray the family (these students would know a little about the family history); assign other members to be the genetic counselors who are responsible for gathering information by asking the right questions of the family members. Have the genetic counselors first brainstorm how to counsel the family and then engage in a role-play in which they conduct a team counseling session with the family.

- Clustering: This is a more structured form of brainstorming in which you begin by writing a word or concept in the middle of the board (e.g., "risk," "client anger," "confrontation"). Using this word as a focus, students brainstorm associated words, phrases, and ideas that you write around the word in a circle of related ideas. You could follow this activity with a brainstorming or a laundry list exercise, in which students generate all of the questions they have about the concept.

Grading and Evaluation

General Criteria

- Make your evaluation criteria clear. For example, we tell students that our evaluation of assignments is based on the quality of information, coherence, consistency, and degree of self-reflection. It is not based on whether we agree with a student's opinions or, for instance, on whether or not we like her or his motives for being a genetic counselor. We also stress that we are looking for improvement in basic helping skills by the end of the course and encourage students to take risks and make mistakes in order to develop their skills.

- Provide samples of good assignments (more feasible once you've taught the course once) that you have obtained from other students—with their written permission. Always remove the student's name and any identifying information from a sample assignment.

- If you assign points, state the number of points the student received at the end of the assignment and explain why points were deducted. Additionally, give one or two general reasons for the grade, especially if you've assigned less than full credit.

- Provide behavioral feedback and try to balance positive and corrective comments. Suggest how the student might improve the next assignment. (Decide if you will allow students to revise and resubmit assignments.) Raise rhetorical questions throughout an assignment to encourage further reflection.

- Remind students to address all aspects of the assignment. We spell out each part of an assignment and state the maximum number of points that can be obtained for each section.

- Although the emphasis of this manual is basic helping skills, during skills practice you will simultaneously need to evaluate and correct *technical* or *content* errors regarding genetic conditions, information, etc. Make it clear to students that on more objective tasks (e.g., calculating a risk rate), accuracy is important. For assignments requiring self-reflection (e.g., personal motives, philosophy of genetic counseling, etc.), look for evidence that students have personalized their responses and aren't just quoting others' ideas. Encourage students to provide specific examples from their own experience.

- Suggest to students that they read their answers aloud in exercises that require them to formulate actual counselor responses. We don't speak the way that we write. Saying them aloud will help students formulate more natural responses.

- Evaluate frequently. This provides multiple pieces of data, gets students comfortable with the process, allows you and your students to assess how they are doing, stimulates ideas for rectifying problems, and prompts students to keep up with the material.

- Give students opportunities to tell you what they think they've learned. For instance, use journaling, self-reflection papers, and *1-minute papers* (Davis et al, 1983). For a 1-minute paper, at the end of a class have students anonymously write for 1 minute about the one or two most important things they learned that day and one or two things they are confused about. Collect and review their papers and clarify any confusion in a subsequent class period.

- Try to get a feel for the overall quality of an assignment when determining the final grade.

- As you get to know students, grade with their backgrounds in mind. For instance, we expect more mature and complete products from students who are older and have previous experience in a human service profession; we recognize that students for whom English is a second language may have some grammatical and spelling difficulties; etc.

Evaluating Role-Plays

Methods for evaluating helping skills are highly subjective. Nevertheless, it is important to be as clear and consistent as you can in evaluating student performance. In some chapters we provide general criteria for evaluating helping skills. You might also wish to develop a standard form or checklist for observers. For instance, you could include the dimensions suggested by Barkham (1988):

- Type of behavior (primary empathy, open question, etc.);
- Skillfulness of the genetic counselor (timeliness, plausibility, relevance, inappropriateness—discrepant from client's viewpoint, etc.);
- Interpersonal manner (empathic, respectful, distant, mechanical, etc.).

Observers can check off any categories that the counselor demonstrates, or they can rate the degree of effectiveness (e.g., poor, adequate, good, excellent).

In our course we stress that behavioral feedback is more important than checks on a rating form. Students learn by hearing specific examples of what they did (e.g., "When you said that the client might be a little nervous about talking to you, she seemed to relax"). They also learn by receiving specific suggestions as to how they might improve (e.g., "Try asking one question at a time so that the client doesn't get too overwhelmed").

Development or Deficiency?

Research has demonstrated that systematic training can improve helping skills. For example, individuals can be taught to formulate more efficient primary empathy responses. However, the *timing* and *choosing* of responses are more advanced skills that will develop gradually as students gain supervised counseling experience (Goodyear and Bernard, 1998; McCarthy and LeRoy, 1998). It also appears to be the case that the "rich will get richer." In other words, students who have adequate levels of cognitive development and who possess enough self-awareness and interpersonal sensitivity to choose appropriate responses in a given situation will tend to become more effective helpers. One challenge you will face is evaluating whether poor performance is due to developmental issues (e.g., lack of experience, naivete) or deficiencies (e.g., lack of self-awareness). Criteria for making this determination include four identified by Lamb and his colleagues (1987):

- The problem behavior is pervasive.
- The student does not acknowledge, understand, or try to do anything about the behavior when it is identified.
- The problem behavior does not improve with training, feedback, or other remediation efforts.
- The student's problematic behavior requires a disproportionate amount of instructor time.

We would add a fifth criterion:

- The student responds defensively to critical feedback by denying or projecting (e.g., "It's your fault I haven't done better"); defending the behavior as a difference in learning style, interpersonal style, or cultural difference when this is not the case; pleading or bargaining; challenging the validity of the helping skills approach; and/or avoiding further feedback.

Student Resistance

It is natural for students to feel varying amounts of anxiety and resistance when learning genetic counseling skills. They may be resistant to different aspects of helping skills training for one or more reasons:

- They are afraid of the unknown.
- They are worried that they are not/will not be good genetic counselors.
- They do not see the relevance of a particular topic or activity.
- They don't want to look foolish or incompetent in front of others.
- They are used to being A students who, in many courses, could memorize material and receive a perfect grade. Helping skills training is quite different because there is always something the counselor could have done differently.
- Your instructions/expectations are not clear and/or are inconsistent.

Furthermore, when students go through basic helping skills training without having ever done genetic counseling, they may tend to (1) think that genetic counseling is easier to do than it is; (2) discount some of the feedback they receive because it's not from a client or clinical supervisor; and (3) discredit some aspects of the helping skills model, theory, and skills. For instance, they may be dogmatic about how to do genetic counseling; this is common for novices who need some certainty in order to deal with their anxiety about being beginners.

Strategies for Addressing Student Resistance

There are several things you can try to work through student resistance:

1. Ask yourself whether the resistance is justified (e.g., you did not provide clear instructions; the relevance of a particular activity is questionable, etc.).

2. Create situations where you gradually increase the difficulty/threat level of what you ask students to do so they can ease into these activities and be more successful in doing them.

3. Provide a rationale for each topic and activity. If you are challenged (e.g., "Why are we doing this anyway? It seems like a waste of time!"), ask the group, "Why do you think we're doing this? What's the point?" Usually someone will come up with a compelling rationale.

4. Point out the resistance and talk about how it's natural to feel a little embarrassed or hesitant when trying out new things.

5. Self-disclose about your own discomfort with certain helping skills and talk about some of the mistakes you have made as a genetic counselor, especially recent ones. This will help students realize that genetic counseling is not easy, and that everyone at any point in training and development has something to learn.

6. If feasible, have volunteers come in to serve as clients for the students during role-plays. Outsiders can provide influential feedback about what worked and what didn't during the role-play. It is more difficult to discount a volunteer's feedback than a classmate's comments.

7. Talk with students about the difference between helping skills training and their other courses with respect to performance and evaluation. Brainstorm with them ways to manage the stress and anxiety of being evaluated on their interpersonal skills and sensitivity (see Appendix 1.2 for guidelines on giving and receiving feedback).

8. At times, you may be the only one who has a realistic or valid perception of what happened during a role-play or other activity and how well or poorly a student did during the activity. You will need to be tactful (especially when the rest of the group says that a student did a wonderful job). You could try saying, "Here's another way to consider doing it..." Or "Let's talk about the plusses and minuses of how you responded when the client said..."

Skills Integration

Because the basic skills are taught one chapter at a time, students may have difficulty seeing how they all fit together. You need to help them understand how to integrate the skills into a model of helping (e.g., demonstrating a complete genetic counseling session). To assist in this process, we offer three activities that require integration. Appendix 1.3 outlines a genetic counseling session interview analysis that can be assigned later in the course when students have covered most of the basic helping skills. Appendix 1.4 lists questions that stimulate students to reflect upon their learning in the course; we use this activity during the last class session. Appendix 1.5 describes an exercise that requires students to articulate their model of helping; it is our final course exercise.

Closing Comments

Remember that teaching and supervising genetic counseling students are skills that improve with practice and experience. At their best, training and supervision are continuous learning processes for both teachers/supervisors and their students. Also keep in mind that helping is as much an art as it is a science.

Every genetic counselor has a highly personalized style of helping. Encourage students to talk with their clinical supervisors about why different genetic counselors in different settings approach genetic counseling differently. Reflection upon these discrepancies will assist students in developing their own individual styles.

Appendix 1.1: Guidelines for Student Role-Plays

A significant amount of class time is spent doing role-plays. Despite their artificial nature, role-plays have been found to be highly effective in increasing trainee-helping skills.

Format

Role-playing will typically be done in a "triad," consisting of three students. When feasible, one co-instructor will sit in on the triad. Students will take turns acting as counselor, client, and observer. Triads will be rearranged frequently so that you get the opportunity to work with different people.

Length

Early in the course, role-plays will be brief—5 to 10 minutes. They will be longer as the course progresses and new skills are introduced.

Roles

1. Counselors are asked to demonstrate the skills covered up to that point in the course.
2. Clients are expected to convincingly present the client's concerns or problems.
3. Observers are expected to watch the role-play and provide the counselor with feedback about the behaviors s/he used well (positive feedback) and about the behaviors that could be improved (corrective feedback). It is strongly recommended that observers make a written record of the role-play so that feedback is specific and chronological. One method is to divide a piece of paper into three sections:

Counselor (Co)	Client (Cl)	Observer (Obs)

In the counselor and client columns, write down key phrases or sentences. In the observer column note any comments you might have about the role-play.

Typical Student Questions About Role-Plays

1. "Should I use any real material when I'm the client?" There is no definitive answer to this question. Real material has the benefit of being closer to an actual genetic counseling situation, and you don't have to struggle, making up information. On the other hand, you risk being caught off guard, revealing more than you intended, or you might feel frustrated when the role-play is cut short and feedback emphasizes the counselor's behaviors. Hypothetical material has the benefit of being less emotionally charged and also allows for a greater variety of situations and client types.

2. "Is it normal to feel anxious during role-plays?" Yes! But you will find that your anxiety decreases as you get more accustomed to doing role-plays. It may also help to remind yourself that there is no "perfect" genetic counseling session and no "terminal skill level." Counselors at all levels of experience can benefit from practice and feedback.

Some Ground Rules for Role-Plays

1. Think through the client role before you begin, so that you are able to respond to the counselor as naturally as possible (e.g., What thoughts, feelings, and behaviors might this client have?).

2. Do not discuss the role-plays outside of your triad or class. Whether hypothetical or real, clients and counselors deserve confidentiality.

3. Think about whether there are certain skills or behaviors you would like the members of your triad to watch for when you are the counselor, and tell them this before a role-play.

4. Realize that you will receive a great deal of feedback from many different individuals during this course. The feedback will vary in its validity and importance. Listen for "themes" across different observers and role-plays. Repetitious feedback is generally a good indicator of your counseling strengths and growth areas.

Appendix 1.2: Giving and Receiving Feedback

Feedback is a way of helping another person to consider changing his or her behavior. It involves communicating your impressions, feelings, and observations about another person's behavior in order to be helpful to the person.

Types of Feedback

There are four types of feedback:

- Positive
- Corrective

- Performance
- Personal

	Positive	Corrective
PERFORMANCE	Reinforcement to Continue Behavior Emphasizes discussing a perception or thought Example: I thought your summary was well organized	Reinforcement to Modify Behavior Emphasizes discussing a perception or thought Example: I think your questions are too lengthy
PERSONAL	Reinforcement to continue a behavior Emphasizes discussing feelings Example: When you listen to me, I really appreciate it	Reinforcement to modify a behavior Emphasizes discussing feelings Example: When you laugh at me, I feel hurt.

Giving Effective Feedback*

Effective feedback is:

- given as promptly as possible after the observed behavior.

- concise; it does not contain unnecessary detail or information.

- focused on the person's observable behavior, not the person's character. For instance, "You didn't look at the client when you talked to her," not "You were strange and distant."

- given in a personal and nonthreatening manner, avoiding moral or value judgments. For example, "When you look away when you talk to the client, I get the impression you are disconnected," not "Nobody likes people who look away when they talk."

- concerned only with behavior the person can modify. For instance, counselors cannot change their gender in order to connect better with a client.

- focused on the person's strengths as well as limitations.

- discussed by both giver and receiver until they can see each other's perspective.

- is definite; not given and then "taken back."

Receiving Feedback Effectively

When an individual receives feedback, the following behaviors can maximize its effectiveness:

- Clarify feedback. Let the person giving you feedback know that you have heard what s/he said. If you do not understand, ask for clarification.

- requested by the receiver, if possible.

*Adapted from Danish et al, 1978.

- Share your reaction. When receiving feedback, you will have both a cognitive and an emotional reaction. Try to be aware of both and decide which aspects you want to share with the feedback giver.

- Accept positive feedback. If feedback is positive (especially when it is information you already know), you may tend to gloss over it quickly. Perhaps you are embarrassed, do not believe positive feedback, feel bored with the same old comments, etc. Try to remember that aware ness of your strengths as well as your weaknesses is important in building strong skills. Try to think of new ways to creatively adapt your strengths in different situations.

- Accept corrective feedback. If feedback is corrective (especially when it is information that is new to you), you may tend to have a negative response. You may deny it, discredit the source, feel threatened or ashamed, etc. Try to remember that corrective feedback is not a global criticism of you as a person (or as a genetic counselor!). Try to identify strategies for correction. You may want to ask for specific suggestions.

- Test validity of feedback. If the feedback does not fit well with your perceptions, pursue it further. You can ask for clarification. You can ask others if their perceptions are the same. While everyone's perception is valid in his or her own frame of reference, it is not necessarily the absolute "truth." Usually you will discover the essential and useful element of feedback if you persist in attempting to understand it.

- Exercise personal responsibility. As you develop as a genetic counselor, you will become more aware of your own strengths and limitations. It is important for you to take responsibility for you own growth early in your development. One way to do this is to ask for feedback about specific behaviors and issues that you know are problematic for you. It can be helpful to ask for this feedback before you are observed. Then observers can focus on these areas as well as other aspects of your counseling behavior.

- Avoid feedback overload. As times you may feel swamped with ideas, and at those times it is OK to let people know that you have had enough for one session. You can always schedule another time to finish receiving feedback. Remember that feedback is for your own growth. If it is coming in too fast to assimilate, it will not help you develop your skills.

Feedback Exercise

This is an exercise that can be used with students early in a course or in a clinical supervision relationship.

- First, ask students to think of a time in their life when they received feedback that was particularly helpful and then to think about what made this feedback so helpful. Next, ask them to think about a time that they received feedback that was not helpful and to reflect upon what made the feedback unhelpful.

Then ask students to share what made their feedback situations helpful and unhelpful, and list these on the board. Add to the list any characteristics they may have overlooked.

- Next, using the definition of feedback presented at the beginning of this appendix, ask students to generate examples of the four types of feedback.

- Finally, ask students to find a partner and discuss this question: "What can you do to help yourself feel safe enough to receive all types of feedback?" Then open a discussion with the whole group about strategies to make feedback easier to give and receive in this course or clinical rotation.

Estimated time: 45 minutes.

Appendix 1.3: Genetic Counseling Session Analysis

Genetic Counseling Session Analysis

The instructor models an entire genetic counseling session with a volunteer from outside of the class while the students observe. The session is taped so that students can watch or listen to it later when preparing the written assignment.

Estimated time: 60 minutes.

PROCESS

Students break into groups of five to discuss what they observed. Estimated time: 30 minutes.

Written Assignment

Based on the counseling session, prepare a seven to nine page, word-processed, double-spaced paper describing the following:

Demographics: A demographic description of the client (e.g., age, gender, ethnicity, general appearance, socioeconomic status, manner of presentation, and motivation).

Client expectations of the session: How did the client describe her/his concerns? Why is s/he seeking genetic counseling?

Medical and family history: Summarize the client's history and provide a risk assessment, if appropriate.

Working conceptualization: What are this client's major (1) medical, (2) genetic, and (3) psychosocial issues? Give evidence to support your conceptualization.

Counseling process: Evaluate the helping relationship the genetic counselor formed. What did s/he do to establish such a relationship? (Hint: Using the types of skills presented in this course, carefully describe what s/he did to help. Include brief statements to illustrate the skills).What goals did the genetic counselor seem to have for this session?What could the genetic counselor have done differently?

Follow-up plan: If the genetic counselor were to meet with this client again, what would be her/his goals for future sessions? What types of additional information are needed? What types of helping skills seem appropriate given the client and the nature of her or his concerns?

Appendix 1.4: Integration of Skills: Stimulus Questions

Dyadic Exercise

Pairs of students take turns answering each of the following questions, or they can select any they wish to answer in the order they wish to answer them. Another variation is to put each question on a separate sheet of paper and students randomly draw out slips and respond to the question on the slip of paper.

1. How would you describe your current level of confidence in your helping skills?
2. What have you learned about yourself as a genetic counselor in training?
3. What do you think is the most important skill?
4. Are there things you are still confused about?
5. What is the hardest skill we've talked about?
6. What is one of the most helpful things you've learned in this course?
7. How has this course affected your communication style?
8. What else do you feel you need to learn?
9. What makes genetic counseling so difficult?
10. What differences do you see between genetic counseling and listening to a friend?
11. What could make you feel "burned out" as a counselor?
12. What is one thing you expected to learn, but didn't? One thing you did not expect to learn, but did?
13. How have your ideas changed about genetic counseling?
14. How helpful is genetic counseling? Why?

PROCESS

Processing in the large group is optional.

Estimated time: 20 to 75 minutes.

Appendix 1.5: Philosophy of Helping Paper

Prepare a six to eight page, word-processed, double-spaced paper in which you present your model of helping clients. You should incorporate readings, classroom activities, and personal experiences. Include in your model a discussion of the following:

1. What is helping? What makes it so difficult? What role does diversity (culture, gender, etc.) play in helping relationships?

2. In this course, we have stressed identifying client concerns, setting goals, and decision-making. Why is each of these activities important?

3. What is the major theory(ies) that underlies your model? Discuss what aspects inform your philosophy and why (see Chapter 2 for a discussion of several theories).

4. What are the major counselor goals, that is, what would a genetic counselor hope to accomplish?

5. What are the major counselor skills (should include behaviors and personal qualities)? Pick one or two skills that you think are most important and explain why. Also discuss how central these one or two skills are to the theory(ies) that underlies your model.

6. What have you learned about yourself as a helper since the beginning of the course? What are your strengths? What are areas for needed improvement? What can you do in the future to improve in these areas? What are behavioral goals that you would like to accomplish to lessen these limitations?

7. How have your beliefs about helping changed since the beginning of the course?

Annotated Bibliography

Bernard, J. M. (1988). Receiving and using supervision. In: Hackney, H., Cormier, L. S., eds. Counseling strategies and interventions (3rd ed.). Englewood Cliffs, NJ: Prentice Hall, 153-169.
[Defines the processes and purposes of clinical supervision and different roles that supervisors and students can use to promote student skill development.]
Cherrin, S. (1993-94). Teaching controversial issues. Teaching excellence, 5(1). Austin, TX: Professional and Organizational Development Network in Higher Education.
[Offers strategies for presenting controversial issues and for managing classroom interactions around these issues.]
Flash, P., Tzenis, C., Waller, A. (1995). Helpfulness of feedback given you about your performance. In: Using student evaluations to increase classroom effectiveness. Minneapolis, MN: Faculty and Teaching Assistant Enrichment Program, University of Minnesota, 58-61.

[Offers practical suggestions for providing feedback and evaluation.]

Hagen, A. S. (1993-94). Learning a lot vs. looking good: A source of anxiety for students. Teaching excellence, 5(2). Austin, TX: Professional and Organizational Development Network in Higher Education.

[Discusses the prevalence of student anxiety, its impact on learning and retention, and how to address anxiety.]

Hill, C. E., O'Brien, K. M. (1999). Helping skills: facilitating exploration, insight, and action. Washington, DC: American Psychological Association.

[Includes rating scales for each basic helping skill. The scales typically range from 1 to 3 and contain behavioral anchors. Instructors and students could use them during student role-plays.]

Johnson, R. T., Johnson, D. W., Smith, K. A. (1990). Cooperative learning: an active learning strategy for the college classroom. Baylor Educator, 15(2), 11-16.

[Provides an overview of cooperative learning with applications to learning settings.]

Johnson, D. W., Johnson, R. T., Smith, K. A. (1991). Active learning: cooperation in the college classroom. Edina, MN: Interaction Book Company.

[Defines active and cooperative learning and provides numerous strategies for using this approach in the classroom setting.]

McCarthy, P., LeRoy, B. S. (1998). Student supervision. In: Baker, D., Schuette, J. L., Uhlmann, W. R., eds. A guide to genetic counseling. New York: John Wiley & Sons, 295-330.

[Offers guidelines and techniques for supervisors to promote student clinical skill development.]

McKeachie, W. (1994). Teaching tips: strategies, research, and theory for college and university teachers (9th ed.). Boston: D. C. Heath.

[Provides a comprehensive review of effective teaching practices, supported by relevant research. Topics include constructing tests, assessment and evaluation, and organizing a course.]

Svinicki, M. (1993-94). What they don't know can hurt them: the role of prior knowledge in learning. Teaching excellence, 5(4). Austin, TX: Professional and Organizational Development Network in Higher Education.

[Discusses how prior knowledge affects subsequent learning and offers practical tips for assessing students' prior knowledge of course content.]

CHAPTER 2

Overview of Genetic Counseling: History of the Profession and Methods of Practice

Learning Objectives

1. Understand the history of the genetic counseling profession.
2. Recognize the major methods of counseling that provide the framework for practice.

History of Genetic Counseling

Genetic counseling as a recognized, independent medical profession is relatively young. However, people have used genetic information for quite a long time throughout history. For instance, the Talmud advises against circumcising brothers of bleeders, and in most cultures throughout history, incest is forbidden. Historically, people made associations by observing patterns of disease in families (Walker, 1998; Weil, 2000). Although there were, and still are, many nonscientific beliefs about the causes of disease, information from such observations was sometimes used to prevent the same problems in future children.

During the turn of the 20th century to the mid-1900s, genetic counseling became the purview of public health and took on the mission of social reform. Heredity was credited as the cause of not only medical conditions but also of many social problems such as poverty, crime, and mental illness. The field of eugenics had dawned, and it was quite a public social movement (Sorenson, 1993). In an early publication, Sorenson (1976) describes this movement as a mission:

It was Arcadian to the extent that many within the movement looked to the past as an ideal and they were attempting to reconstruct an assumed lost purity of the American race, or to recapture the simplicity of an earlier form of social existence. The movement was also Utopian in that it looked to the future as an opportunity to improve men and society, through selective breeding, immigration, and social planning. In both its Arcadian dreams and Utopian fantasies, it looked to genetics as the method. [p. 474]

The general public as well as psychologists, medical professionals, and politicians were eager to use this new science of genetics to improve the human race, believing that this goal would benefit everyone. In the early 1900s, many states in the United States had laws mandating the sterilization of the mentally defective. Informed consent was not involved, and thousands of people were sterilized involuntarily. Inferior ethnic groups were not allowed to immigrate, and by the 1930s these ideas were well accepted not only in the U.S. but also in many other countries as well. In Germany these ideas became the horrible excuse for killing many people considered to be inferior. This part of the history of genetics is important to remember because it colors the public perception of the profession of genetic counseling. Today many people are still reluctant to seek genetic counseling, fearing that they will be told not to have a child because of their "bad" family history. Additionally, the potential for genetic discrimination has become a real risk, and the fears of both patients and providers about this risk appear to have some validity (Harper, 1993; Murray, 1992).

In 1947, Dr. Sheldon C. Reed coined the term *genetic counseling*. He delineated the three requirements of genetic counseling: (1) knowledge of human genetics; (2) respect for sensitivities, attitudes, and reactions of clients; and (3) teach and provide genetic information to the full extent known (Reed, 1955). This era of genetics marks a significant turn in that the scientific community took over the ownership of the practice of genetic counseling and adopted the belief that individuals should make decisions for themselves about their genetic risk. In commenting in the Preface on the first edition of the very first book that was published on the practice of genetic counseling, *Counseling in Medical Genetics*, Reed (1980) writes, "The first edition was written as an introduction for physicians to the new subject of counseling in medical genetics. It was my hope that it would have a wide distribution, and it did. Thousands of physicians enjoyed the comic bits in it, hopefully some learned a little genetics, and all were introduced to genetic counseling".

In 1975, the American Society of Human Genetics (ASHG) published a definition of genetic counseling (ASHG, 1975). The most important aspects of the definition include the acknowledgment that genetic counseling is a communication process and that patient autonomy is the guiding principle of that process. In the third edition of his book, Reed acknowledges the formal definition of genetic counseling from the ASHG but defines it for himself as "a kind of social work which is often medical but not always so" (Reed, 1980, p. 9). These definitions form the framework for contemporary genetic counseling practice, and the underlying principles are important in that they distance the field from eugenics, and at the same time, strive to empower the client. Patient autonomy is valued over any other factor. Over time the definition of genetic counseling will likely change as the profession continues to grow in response to new challenges and opportunities.

Dr. Seymour Kessler (1980) described the major change in emphasis in genetic counseling as a *paradigm shift*. He was talking about a shift that originally emphasized a eugenics model to a preventive medicine model and then to a psychosocial medicine approach that emphasized patient self-determination and the genetic counselor's role as a patient advocate, grief counselor, researcher, and health care professional providing supportive care, education, resources, and referrals (Kessler, 1980). This shift has positioned the practice of genetic counseling to be a unique, multifaceted health care service.

As heredity clinics sprang up at major medical centers in the United States, genetics was well on its way to becoming an established medical service. In the third edition of his book, Reed (1980) also talks about a potential demand for genetic counseling that will exceed the supply of genetic counselors. He was referring mostly to physicians and Ph.D. geneticists at that time. The practice of genetic counseling as it exists today was just beginning to develop, but his prediction proved to be right on target.

In 1969, at Sarah Lawrence College in New York, the first graduate program designed to educate health care professionals specifically to provide the service of genetic counseling enrolled its first class of students. The curriculum evolved over the first few years to incorporate the study of the psychosocial dimensions of genetic counseling with the medical aspects of genetic disease (Marks, 1993). The profession was destined to be a hybrid, drawing skills from both the medical and the counseling professions. This graduate program, and others that soon followed, educates students to be unique health care professionals who help their clients cope with both the medical and the psychosocial aspects of genetic risk and disease. These new genetic counselors saw their responsibilities as not only involving the provision of genetic risk information but also working with families to help them understand their condition and their options for dealing with the condition, facilitating decision making, and providing psychosocial supportive services (Eunpu, 1997). At the time that this chapter is written, there are approximately 2000 practicing genetic counselors in North America, and growth in the profession is adding about 200 new professionals each year. Additionally, 27 graduate genetic counseling programs have been established in North America, and programs are in existence or under development in the United Kingdom, Australia, and South Africa. The National Society of Genetic Counselors (NSGC) Web site presents a picture of the scope of practice in 2001:

Genetic counselors are health professionals with specialized graduate degrees and experience in the areas of medical genetics and counseling. Most enter the field from a variety of disciplines, including biology, genetics, nursing, psychology, public health and social work. Genetic counselors work as members of a health care team, providing information and support to families who have members with birth defects or genetic disorders and to families who may be at risk for a variety of inherited conditions. They identify families at risk, investigate the problem present in the family, interpret information about the disorder, analyze inheritance patterns, and risks of recurrence and review available options with the family. Genetic counselors also provide supportive counseling to families, serve as patient advocates and refer individuals and families to community or state support services. They serve as educators and resource people for other health care professionals and for the general public. Some counselors also work in administrative capacities. Many engage in research activities related to the field of medical genetics and genetic counseling.

Today's genetic counselors help thousands of clients each year either through direct care or indirectly through their work in the research and commercial arenas. They work in a wide variety of settings and possess a multiplicity of skills from basic science to counseling, teaching, research, management, and more. Genetic counselors have taken, and will continue to take, the profession in new and different directions. Nevertheless, strong communication skills will always serve as the basic framework that sets the foundation for genetic counseling.

Models and Methods of Practice in Genetic Counseling

Genetic counseling is directly concerned with human behavior. Thus it must be based on a knowledgeable understanding of psychodynamics and of the principles of interpersonal functioning. Also needed is an understanding of the psychological meanings of the issues with which genetic counseling is involved, namely the issues of health-illness, procreation, parenthood, as well as the complex processes by which the goals of genetic counseling are achieved (Kessler, 1979, p. 21)

So, what is the model of practice for genetic counseling? This is a question that many have grappled with for quite some time, and there is still no good answer. In examining the evolution of the nondirectiveness approach to genetic counseling (more on that later), Fine (1993, p. 107) states, "Just as the practice of genetic counseling incorporates skills and knowledge from many disciplines, so have the philosophy and principles evolved in an eclectic way." The profession has drawn upon multiple psychological models and methods as the practice of genetic counseling has developed over the years and continues to evolve. Djurdjinovic (1998, p. 127) discusses her personal struggle "to understand what genetic counseling was all about." She clearly describes the two distinctly different elements that entice genetic counselors into the profession. Genetic counselors are excited by the challenge of having to constantly learn about the technical and medical aspects of genetic disease. Not only is it interesting, it is constantly changing. At the same time, they are driven by the need to understand the human meaning of the impact of genetic disease. Djurdjinovic sums this up by stating, "Almost all genetic counseling sessions contain competing priorities between the medical and psychological considerations of a genetic condition" (p. 127).

It is obvious that no one counseling model or method corresponds very well to the process of genetic counseling and that is likely due to the unique nature of the practice. Kessler challenges us time and again to evaluate the specific and distinctive processes of genetic counseling in order to describe the model and method that is unique to the field and then use this model to evaluate and improve effectiveness. "The paucity of process studies hinders us from finding ways to make the educational and counseling aspects of genetic counseling more effective" (Kessler, 1999, p. 342).

Since the primary focus of this manual is on basic helping skills, this chapter concentrates on some of the more basic models and methods that have been applied to the profession. It is important to note that many other models and methods have been proposed, some quite complex, but they are beyond the scope of this manual.

Prior to writing this text, Dr. Sue Petzel of the University of Minnesota conducted an informal survey of graduate genetic counseling programs in North America to assess which psychological models are perceived to be very important to the practice of genetic counseling. Petzel, a clinical psychologist who works with patients and families with genetic conditions, gathered this information to prepare her lecture materials for teaching genetic counseling students in training. Fourteen programs responded to the survey, indicating the model(s) they considered to be the major counseling emphasis in their curriculum. Nine programs reported that the client-centered model was the major counseling emphasis; two endorsed the psychodynamic model, two the behavioral model, one the developmental model, and two the family systems model (there are 16 answers from 14 programs because one program rated the client-centered, psychodynamic, and behavioral models as equally important). It should be noted that all but two of the programs endorsing models other than client centered as a major emphasis also rated client centered as second choice. It is clear that the client-centered model is a major component of the counseling aspect of genetic counseling practice today.

To understand why the profession has embraced Carl Rogers' client-centered counseling (which he later renamed person-centered counseling) as the dominant counseling approach, one needs to think about the underlying values of the profession and the major goals of practice. The professional code of ethics for genetic counselors illustrates the basis for this approach by emphasizing respect for client's beliefs, background and culture and the counselor's duty to enable clients to make autonomous decisions by providing all necessary information (Benkendorf et al, 1992). Genetic counseling uses many of the same methods and frameworks of the theoretical models of mental health counseling and psychotherapy, but traditional genetic counseling is not meant to be psychotherapy. Genetic counseling has different goals from traditional psychotherapy. Genetic counseling strives to empower patients through education and to facilitate autonomous decision-making (Fine, 1993). The approach of client-centered counseling that Rogers described forms the groundwork because it appears to best meet the basic values and goals of the profession.

Rogers' Person-Centered Counseling

The basic philosophy fundamental to Rogers' theory involves a positive view of humans and a trust in the self for greater inner directedness. Rogers believed that people possess the capacity to become self-aware, to self-direct, and to actualize into whole, fully functioning individuals (Rogers, 1992). The aim of therapy in Rogers' view is to help clients in their personal growth process so that they are better able to cope with the difficulties they are facing now as well as those that occur in the future (Corey, 1996).

Rogers' basic assumptions about human nature and psychosocial development have been described as follows (Hjelle and Ziegler, 1984):

- People are inherently free to *self-actualize*. They are able to overcome conditions of worth (other's expectations) to make the choices that help them develop into the persons that they become.

- People are basically rational beings—planful, thoughtful, etc. Irrational behaviors stem from being out of touch with one's true inner nature.

- The human self is global, all-inclusive, and whole, not compartmentalized. Therefore, one must try to understand a person in her or his entirety.

- People have the innate potential to self-actualize, but this potential can be compromised by environmental events. Nevertheless, individuals can rise above such events.

- A person's essence is her or his self-concept or self-perception. The private world of experience shapes the self-concept.

- The actualizing tendency is purposeful and oriented toward the future. People grow from, rather than react to, external experience.

- The actualizing tendency moves an individual toward growth, self-realization, and personality enhancement.

- The actualizing tendency itself leads to constant growth and unfolding potential.

- No one can fully understand another's private world.

These assumptions form the foundation of person-centered counseling. For instance, Rogers theorized that the attitudes of the counselor and the quality of the counselor-client relationship determine the outcome of the counseling process. It is important to note that the major focus is on the counselor's way of thinking. Rogers believed that when the counselor is able to converge key elements, a positive outcome would follow. He described three key counselor attitudes that he referred to as facilitative conditions:

- Unconditional positive regard: a positive view of clients, the belief that clients are coping to the best of their ability, a total respect for who they are as individuals, acceptance of client strengths and weaknesses, a belief that clients have the capacity for self direction, and a focus on the present moment and experience.

- Empathy: strives to understand the client's reality, to get into the world of the client, and to see things from the client's perspective.

- Counselor genuineness: the counselor establishes an open relationship with the client, is open to his or her own emotional reactions to the client, and establishes a safe setting where clients are free to self-actualize.

Upon examination of Rogers' theory, one can see how the basic foundations of this approach are a good fit with the values and major goals of genetic counseling. This approach empowers the client, values the client's belief systems, and strives to understand the client's experiences. The combination of respect for autonomy and the application of Rogers' theory and practice to reproductive decision making form the basis for the *nondirective* tenet in genetic counseling (Fine, 1993).

What exactly is the nondirective approach? One might question if it is an approach at all. Kessler (1997c) describes it as, "a way of thinking about the relationship between client and counselor." Bartels et al (1997) surveyed genetic counselors in the attempt to describe nondirectiveness in actual practice and found that there was general agreement with respect to the definition of nondirectiveness. Most respondents provided definitions that include the intention of enabling clients to make independent, informed decisions free from coercion. However, respondents were more variable in their descriptions of *how* they accomplish this nondirectiveness goal. Additionally, it appears that genetic counselors recognize that they direct the *process* of genetic counseling but are careful not to direct the *outcome* (Bartels et al, 1997). It is our opinion that nondirectiveness serves as a good reminder that as genetic counselors we must take care to be aware of our own biases and intentions and strive to understand the issues of the session from the client's perspective. However, nondirectiveness does not fully capture the counselor's behaviors or the genetic counseling process.

We need to look at the work of Kessler when considering other potential models and methods to apply to genetic counseling. No one individual has examined the psychological dimensions of genetic counseling in more detail than has Kessler. His work in this area reaches back to the time when genetic counseling was first establishing itself as a health care service in major medical settings. He has been tireless in his efforts to examine the psychological issues associated with the provision of this service and to teach genetic counselors about the skills they need to enhance their practice as the profession has advanced over the years.

Person Oriented Counseling

Kessler (1979) discusses the emergence of a more psychologically or person-oriented genetic counseling in contrast to the content-oriented approach. He states that the person oriented approach

starts with the premise that genetic counseling deals with human behaviors, important ones at that; health and illness, procreation, parenthood, and sometimes life and death. It views the problems posed by a genetic disorder as being intimately related to the overall situation of the persons, their ways of solving problems, making decisions and adapting to life crises. Whereas the content oriented approach emphasized facts, the person oriented approach places the focus on the various meanings that facts have for the counselees as well as on the intrapsychic and interpersonal consequences of these meanings. [p. 19].

In the content-oriented approach:

- The counselor believes that objective facts and figures are the basis on which decisions are made and actions proceed.
- The counselor gives high priority to providing information.
- The counselor functions as authority, educator, and advisor.
- The approach fosters emotional distancing on the part of the counselor.

In the psychologically oriented approach:

- The counselor believes that decisions and actions are based on the subjective understanding and varied meanings of the facts and figures.
- The counselor helps clients understand and integrate their experiences.
- The counselor functions as facilitator, guide, and model.
- The approach fosters counselor involvement with the client's emotional issues.

The Teaching Versus Counseling Approach to Genetic Counseling

Kessler (1997a) describes two basic approaches to genetic counseling: teaching and counseling. These models dovetail well with the descriptions of the two basic orientations of the genetic counselor that Kessler described 18 years earlier.

THE TEACHING MODEL OF GENETIC COUNSELING

- The major outcome goal is educated clients.
- A premise is that clients come to genetic counseling for information.
- An assumption is that informed clients are able to make autonomous decisions.
- Cognitive and rational processes form the foundation of the approach; psychological aspects are minimized.
- The counseling process involves providing all-inclusive, accurate information in an impartial manner; the counselor does not become involved.
- Teaching is the only means to meet the end goal: an educated client.
- The counselor-client relationship is based on counselor authority.

THE COUNSELING MODEL OF GENETIC COUNSELING

- The major outcome goals are to understand the client, advance the client's sense of self-competence, help the client gain a sense of control, alleviate some psychological stress, provide support, and help the client with problem solving.
- A premise is that clients come to genetic counseling for complicated reasons such as needing information, wishing for validation, wanting support, and looking for a way to reduce their anxiety.

- Human behavior and psychological aspects of genetic counseling are complex.
- The counseling process is multifaceted, involving the psychological assessment of client strengths, limitations, needs, values, and decision making styles; a range of counseling skills are needed for a positive outcome; counseling must be specific to the client and flexible; and the counselor must attend to his or her inner self.
- Education is only one means that is used to meet the end goals described above.
- The counselor-client relationship is mutual.

In comparing the models, Kessler (1997a) states, "The net psychological impact of this strategy [the teaching model] is to enrich the authority, status, and ego of the professional at the expense of the client. The purpose of any counseling strategy is to reverse this process and leave the client psychologically enriched even if it is at the expense of the professional." Maturation into the profession requires students to grow away from being the content-oriented counselor utilizing the teaching model into the psychologically oriented counselor who makes use of the counseling model.

Counselors strive to perfect their ability to understand clients, give them a sense of being understood, and help them feel more hopeful, more valued, and more capable of dealing with their life problems. Because genetic counselors work with people filled with uncertainty, fear of the future, anguish, and a sense of personal failure, they have unusual opportunities to accomplish these tasks. [Kessler, 1997a, p. 341].

Family Systems Counseling

The application of family systems theory to genetic counseling has been suggested because dealing with genetic issues means dealing with family issues. Even if the client is the only affected person in the family, family values, religious beliefs, myths, secrets, etc., will play a role in how the client perceives the disorder, deals with it, and makes decisions. The genetic counselor needs to demonstrate a respect for the client's family beliefs, values, and perceptions in order to build a trusting relationship (Weil, 2000). Weil (2000) describes several aspects of family beliefs that he views as especially pertinent to the practice of genetic counseling:

Beliefs concerning the normal and acceptable range of human variability: In working with families it becomes clear that families perceive the burden of diseases differently. What is acceptable for one family is not for another. These perceptions will play a role in determining an individual's concerns, the testing that s/he is interested in having, the impact of the test results, and what the individual will do with the results.

The personal and social meaning of the particular situation or disorder: This refers to the attributes that a family perceives as important (or unimportant, for that matter). One family might value intelligence over any other characteristic, while another may value physical abilities. These values will play a role in perceptions of burden or acceptance of a particular disorder.

Beliefs concerning the ability to affect the course of life events: This refers to the extent to which a family believes that it can or should do something to change its circumstances. This involves emotional reactions stemming from core family values.

Beliefs concerning the cause of genetic disorders, birth defects, and other situations relevant to genetic counseling: This refers to family myths and religious, superstitious, and medical beliefs, such as thinking that certain prenatal exposures cause a birth defect. In addition, it takes into account feelings of guilt, shame, and punishment that family members have taken on in the face of these conditions.

Weil (2000) argues, "Thoughtful questions concerning beliefs about the cause of the situation that led to genetic counseling will help to identify misinformation, misunderstanding, disagreement among family members, beliefs that are contrary to the medical genetic explanation, and issues that are of particular concern or are sources of guilt or blame" (p. 36).

Intersystems Counseling Model

Eunpu (1997) proposes an intersystems model as being very applicable to genetic counseling as it addresses the individual, interactional, and intergenerational issues intrinsic to genetic disorders. The basic principle of this model is that counseling must include the client's individual psychological framework, the interactions of the couple (client and partner), as well as the intergenerational influences of the client's family. She suggests that genograms could be used by genetic counselors to better understand, and to help the client better understand, the beliefs and values within a family and the impact. Eunpu presents some examples of areas to explore with questions that can be used to help clients understand the roots of their perceptions about the issue at hand for genetic counseling.

Example of areas to be explored with a client who comes to counseling after a pregnancy loss:

- The expectations of the family for the client regarding reproduction (how many children, when to have them, etc.)
- The family messages regarding importance of children.
- The family messages regarding abortion and contraception.
- Any conflicts between the client's family beliefs and those of the client's community, culture or religion.
- The messages, myths, and beliefs the client heard about the pregnancy loss.
- The experiences with pregnancy loss or childhood death in the family.
- The meaning to the client of not being able to meet family expectations

regarding these issues.

Example of areas to be explored with a client who comes to counseling because of a genetic disease in the family:

- Identify the family member(s) thought by the family to be affected.
- Identify the family members who think of themselves as being affected.
- The family messages regarding the disorder.
- The attitudes of the family members who are affected.
- The impact of the disorder on the family dynamics.
- The meaning for the client to think or act differently than other family members regarding the disorder.

This previous example could be most useful in working with clients who come to genetic counseling for predictive testing such as with Huntington disease or a familial cancer. Exploring these issues with the client will help build a trusting relationship, better clarify the client's decisions about testing, and help the client deal with the results.

Psychodynamic Model

This is the model of psychotherapy most associated with Freud and often employed by psychiatrics when treating a long-standing chronic complaint or illness. Freud's belief of human nature is that it is basically deterministic. Human behavior is determined by irrational forces, unconscious motivations, and drives that evolve through key psychosexual stages in the first 6 years of life (Corey, 1996). It was Freud's conviction that psychotherapy imparts insight, and with insight one is able to bring the unconscious into the conscious, thereby freeing one's self to make choices about behavior. Some of the basic concepts are as follows:

- All behavior is determined. Nothing happens at random, even events such as dreams, slips of the tongue, failures in memory, etc.
- The personality develops sequentially through stages. Events that occur early in life have tremendous significance on personality development and therefore present behavior.
- Personality consists of three systems: the id, the ego, and the superego. The id represents the biological component (what we are born with); the ego represents the psychological component (the reality of the outside world, realistic thinking, and intelligence); and the superego represents the social component (internalization of personal values and social standards).
- Life instincts are central to the survival of the individual and they are oriented toward growth, development, and creativity. Behavior is motivated by sexual and aggressive instincts comprised of psychic energy.
- Conflict is, in general, between these instincts (id) versus internalized parental

and social values (superego) and reality (ego).

- All behavior problems have their origin in conflict. They are either direct or indirect expressions of impulses (instincts), defenses against the impulses or breakdowns of defenses.

This approach is intended to give the counselor a comprehensive insight into the client's complaint or illness in relation to his or her development from a psychological perspective. It takes much time but the client can discover significant connections to events that occurred early in life and learn how to change unwanted behavior. Because of the long-term nature of this approach, it is limited in utility in the genetic counseling sphere but it can provide a framework for understanding human behavior.

Cross-Cultural Counseling Issues Important to Genetic Counseling

When we examine elements that compose who we are, culture is one of the most significant ingredients. Cultural influences take into account our families, our histories, traditions, values, and many other facets of our being that are often not conscious. To ignore cultural influences is to ignore the framework of an individual's way of thinking and behaving. If genetic counseling is to be effective, it is vitally important to pay attention to cultural differences between the counselor and the client and focus on those that may be impacting the behavior of the client.

Sue and Sue (1990) emphasize that the social and cultural framework from which the counselor and the client were created influences traditional counseling and psychotherapy. In the United States that framework is almost exclusively white and Western European. This is also true for the traditional approach to medicine in the U.S. Since the practice of genetic counseling is based on the traditional models of counseling and medicine, one can assume that variations in cultural perspective have been overlooked in the provision of this service as well.

Sue and Sue (1990) state that when working with clients whose culture is different from your own, it is necessary to do the following:

1. Be aware of the sociopolitical forces that have impacted the client.
2. Understand that culture, class, and language factors can act as barriers to effective cross-cultural counseling.
3. Point out how expertness, trustworthiness, and lack of similarity influence the client's receptivity to change and influence.
4. Emphasize the importance of worldviews and cultural identity in the counseling process.
5. Understand culture-bound and communication style differences among various cultural groups.
6. Become aware of one's own cultural biases and attitudes.

What are the characteristics that comprise a culturally skilled counselor? Like other counseling skills, these have no terminal skill level. Becoming culturally competent is a continuing learning process. Also, as with other counseling skills, our clients have much to teach us in this process. Sue and Sue (1990) stress three major goals to work toward becoming a culturally skilled counselor:

- Actively engage in the process of becoming aware of one's own assumptions about human behavior, values, biases, preconceived notions, personal limitations, and so forth.
- Actively attempt to understand the worldview of one's own culturally different client. In other words, what are the client's values and assumptions about human behavior, biases, and so on?
- Actively engage in the process of developing and practicing appropriate, relevant, and sensitive intervention strategies and skills in working with culturally different clients.

As with other aspects of counseling, there is no one correct way to counsel in a culturally competent manner. Vivian Wang has done a great deal of work in the area of cross-cultural counseling with respect to the practice of genetic counseling, and is uniquely trained in both genetic counseling and counseling psychology. Her handbook, *The Handbook of Cross-Cultural Genetic Counseling* (1993), offers a systematic way of learning about these issues. She states that genetic counselors must develop a foundation of cultural knowledge, awareness of self and others, as well as specific counseling skills. She offers the following guidelines for providing culturally relevant genetic counseling:

1. Genetic counselors must understand the assumptions made by both themselves and the client regarding the presenting question.
2. Genetic counselors must be aware and understand their own worldviews and the impact of worldview in clinical practice.
3. Genetic counselors must develop a sensitive perception, understanding, and knowledge of their own health care belief system and how it impacts the value system of the client.
4. Genetic counselors must understand the client's history of adaptation and adjustment processes in the United States.
5. Knowing the geographic and temporal distance of the country of origin and the duration of residency in the United States of clients and their families may help facilitate understanding and awareness of the client's degree of alliance to a cultural group membership and or American cultural norms.
6. Genetic counselors must understand and have the knowledge and skills to know how and when to integrate Western genetic testing and treatment and culturally specific interventions.

7. Genetic counselors must be knowledgeable about cultural communication patterns.

8. Genetic counselors must respect the choice of language of the client. Genetic counseling must be provided in the language preference of the client through the counselor, interpreters, and/or genetic assistants.

9. Genetic counselors must be sensitive to and understand the opportunities and restrictions of the institution and society in general that are imposed upon the genetic counseling relationship.

The common thread woven throughout the fabric of cross-cultural counseling is that of respect. Respect implies that there is the desire to understand and accept the differences in individuals. As these are the major facets of the genetic counselor's code of ethics (see Appendix A), it is our obligation as genetic counselors to work toward being competent in every aspect of counseling, including developing skills in the area of cultural competence.

Closing Comments

Genetic counseling models of helping are dynamic. They change over time and with increasing experience. They become more crystalized for individual counselors as counselors grow from novice to expert. They also continue to change more globally as the profession itself grows and expands and new technologies develop.

Today, person-centered counseling remains the keystone of the profession. As the genetic counseling profession evolves, the model is incorporating constructs, methods, and additional perspectives. It is becoming more sophisticated, and in time will prove to be a model recognized to be more distinct and specific to the service. The process of genetic counseling will become a model in its own right.

Annotated Bibliography

Alder, B. (1995). Psychology of health, applications of psychology for health professionals. Luxembourg: Harwood Academic Publishers.
[Provides an overview of various models of health psychology including topics such as health and illness behavior, models of health, and social factors that influence health and others.]
Corey, G. (1996). Theory and practice of counseling and psychotherapy (5th ed.). Pacific Grove, CA: Brooks/Cole.
[Covers the concepts and application of the major models of therapy used in psychotherapy today.]
Fisher, N., ed. (1996). Cultural and ethnic diversity, a guide for genetics professionals. Baltimore, MD: Johns Hopkins University Press.
[Provides an overview of various divers cultures including information about the history, demographics, religion, belief systems, and implications for genetics services.]

Israel, J., ed. (1995). An introduction to deafness: a manual for genetic counselors. Washington, DC: Gallaudet University.
[Presents medical information about hearing and the diagnosis and causes of hearing loss, communication systems, the culture of deafness, the psychosocial aspects of deafness, and the application of the information to genetic counseling.]

Resta, R. G., ed. (2000). Psyche and helix, psychological aspects of genetic counseling. New York: John Wiley & Sons.
[A valuable compilation of essays by Seymour Kessler covering topics such as counseling skills, advanced counseling techniques, directiveness, and psychosocial issues specific to genetic counseling.]

CHAPTER 3

Listening to Clients:
Attending Skills

Learning Objectives

1. Define attending skills and their functions in genetic counseling.
2. Distinguish between physical and psychological attending.
3. Develop attending skills through practice and feedback.

Definition of Attending Skills

Attending is a skill that involves the genetic counselor *observing* client verbal and nonverbal behaviors as one way of understanding what clients are experiencing, and *displaying* effective nonverbal behaviors to clients during genetic counseling sessions. Egan (1994) elaborates upon these two major aspects of attending, which he refers to as "psychological attending," and "physical attending," respectively.

Psychological Attending

Psychological attending is sensing experiences, to the extent possible, through the eyes of the client rather than through your own. You intuit the feelings and attitudes that clients have or might have had by being *in tune* with both verbal and nonverbal messages. Psychological attending involves being sensitive to client feelings and experiences. It consists of both perceiving and processing various client messages. For example, many clients communicate their emotions only nonverbally (Cormier and Cormier, 1991). Therefore, psychological attending is an important counselor skill for recognizing client feelings.

Physical Attending

Physical attending is the way that you use your body to communicate your understanding to the client. Good physical attending can alleviate client apprehension. For example, clients may be fearful or uncertain about what is going to happen in genetic counseling. They may also believe that the genetic counselor is withholding information or is going to tell them what to do. They will attempt to *read* counselor nonverbals for any signs of dishonesty, discomfort, etc. Whereas verbal communication is intermittent, nonverbal communication is constant (Egan, 1994). Furthermore, as much as 93% of the meaning of a message may be perceived from the speaker's nonverbal behaviors (Mehrabian, 1976). Through nonverbal behaviors we "convey what we wish to express and also what we do not intend to express" (Hill and O'Brien, 1999, p. 81). Feelings have a way of working themselves into our nonverbals (e.g., looking down at notes, looking through a medical chart while the client is speaking, looking at your watch, etc.).

Attending is positively related to other counselor skills such as empathy and to counselor attitudes such as respect and involvement (Egan, 1994). For instance, good attending leads to greater empathy and suggests that you are respectfully engaged with your clients. We would go even further and claim that *all* genetic counseling relies on a solid base of attending behaviors. Effective psychological and physical attending can convey involvement and understanding; build rapport; encourage clients to self-disclose; increase client perceptions that you are expert (competent and professional), socially attractive (warm, likable), and trustworthy; and build a foundation from which you may understand and assist clients in decision making.

Attending can be a difficult process for a number of reasons. For instance, "Many social interchanges are marked by a detached passivity in which the 'listener' is just waiting for the other person to stop talking in order to get his or her next speech into the conversation" (Martin, 2000, p. 18). Additionally, you can be distracted, preoccupied with your own concerns, and/or be anxious about your performance. These obstacles must be overcome before you can listen fully to your clients.

Effective Counselor Psychological Attending Skills

Effective psychological attending consists of three major activities: observing and responding to client nonverbal behaviors, understanding client body and facial movements, and attending to subtle cues.

Observing and Responding to Client Nonverbal Behaviors

- Pay attention to client nonverbals (e.g., If the client is gripping the arms of the chair, it may indicate that s/he is anxious).
- Notice incongruities between client nonverbal and verbal behaviors (e.g., saying "yes" while shaking one's head "no"). In general, you should go with the nonverbal behavior when there are incongruities. Nonverbal behaviors are more spontaneous, and therefore more indicative of the client's psychological/ emotional state than are verbal behaviors (Cormier and Cormier, 1991).

- Consider pointing out nonverbals to the client (e.g., "You say that you are fine with this decision, yet you look tearful").
- Comment on nonverbals when a client is silent.
- Look for patterns of behavior that together suggest the client is feeling or thinking a certain way. Beware of overinterpreting a single nonverbal behavior.
- Observe the following client characteristics, as these may give you clues about a client's emotional state, attitudes toward counseling, and motivations: activity level (agitated or lethargic/depressed); manner of dress (sloppy or careful, appropriate to the situation or haphazard); movements (easy or difficult); state of health; tension behaviors (swallowing, nervous laughter, excessive throat-clearing, etc.); voice (firm or shaky; confident or unsure); client projection of self (mature and in control or childlike, submissive or aggressive) (Fine and Glasser, 1996).

Understanding Client Body and Facial Movements

- Attend to the face because it is the richest source of nonverbal communication (Harper et al, 1978). As much as 55% of a client's feeling messages come through the face (Egan, 1994). Over 1000 facial expressions have been identified, and many of these expressions seem to have similar meanings to people from all countries and cultures (Ekman and Friesen, 1969, 1984).
- Facial muscles can be controlled in all areas but the eyes (Hill and O'Brien, 1999). They reveal more of the true self. Watch a client's eyes for signs of fear and anger, which are clearly expressed there. For instance, frightened or anxious people will have dilated pupils, and angry people will have constricted pupils (McSee, 1985). Additionally, anger, distress, and fear can be communicated in the temples (the pulse will quicken); in the carotids (blood visibly pulses to the head); in the upper and lower jaw muscles (clench together); and in the nostrils (dilate and constrict).
- Look for leg and foot movements and physiological reactions. Leg and foot movements are most subject to nonverbal leakage because we move them more automatically, without conscious thought (Ekman and Friesen, 1969). Very controlled individuals often show emotions in their hands and feet (e.g., gripping their hands together, tapping their foot), or in physiological behaviors such as blushing, sweating, breathing (e.g., shallowly or rapidly), blinking, etc.
- Notice frozen expressions (avoidance of showing emotion, poker face); masking (replacing a felt emotion with another more appropriate one); minimizing expressions (to make a feeling seem milder); exaggerating an expression (e.g., nodding the head vigorously and saying "Uh-huh" when one is confused by the information that you present).
- Nonverbal expression may be age- and gender-dependent (e.g., women may be more likely than men to cry).

Attending to Subtle Cues

- Let clients complete sentences.
- Listen for incomplete sentences. Clients may trail off or shift to another sentence/topic because what they were saying is emotionally charged. Ask clients to finish incomplete sentences.
- Try to differentiate between what clients are feeling with regard to the counseling experience as opposed to their feelings about the problem at hand.

Effective Counselor Physical Attending Behaviors

There are five major domains of physical attending: face and eyes; body; voice, distracting behaviors; and touch.

Face and Eyes

Use of occasional head nods

Smiling at appropriate times

Looking at client without staring

Body

Relaxed but alert posture

Hand and arm gestures for emphasis

Legs and feet are still

Oriented toward the client

A comfortable distance from the client

Voice

Adequate volume

Appropriate pace or speed

Use of inflections

Use of words that the client can understand

Tone matches the content and tenor of the conversation

Distracting Behaviors (Behaviors to Avoid)

Excessive use of filler words (e.g., "you know"; "right"; "OK"; "It's like..."; "Um")

Shuffling through medical records

Playing with pen, jewelry, paper clip, etc.

Twisting fingers or hair

Chewing gum

Touch

- Touching clients: Be very careful about touching clients, because your actions could be misinterpreted. Touch can be beneficial if it leads to an improved relationship but harmful if the client views it negatively (e.g., some individuals dislike being touched by strangers; abuse survivors may be particularly sensitive to touch that they did not invite/initiate), or if it is culturally inappropriate. Touch raises a number of social, cultural, ethical, and legal issues.

- Shaking hands: Note that shaking hands is appropriate to Western cultures, but may be regarded as offensive in some cultures (e.g., some individuals from Middle Eastern cultures shake hands only with a person who is of the same gender).

Hill and O'Brien (1999) developed the useful acronym *ENCOURAGES* to organize and describe helper physical and psychological attending behaviors that are generally effective:

E = has moderate levels of *eye contact* (avoids staring or looking away too often)

N = uses a moderate amount of head *nods*

C = respects and maintains awareness of *cultural differences* in attending

O = uses an open stance toward clients (arms open, faces client, legs uncrossed)

U = uses acknowledgments such as "Um-hmm"

R = *relaxes* and is natural

A = *avoids distracting behaviors*

G = matches client *grammatical style*

E = listens with a third *ear* (psychological attending)

S = uses *space* appropriately

Additional Suggestions for Attending Effectively

To convey your understanding to the client in a trustworthy way, we recommend the following:

- Be congruent. Your body should parallel your verbal message. When your words and nonverbal behaviors are discrepant, the client will tend to believe your nonverbals. For example, if you say, "I can respect your decision to terminate the pregnancy" while you are frowning and clenching your hands together, the client will probably decide that you disapprove of this decision.

- Get in sync. Your communication will be more effective if your demeanor is in harmony with your client's. For instance, if your client is very sad and speaking in a slow, barely audible manner, you will have synchronicity if you also slow the pace of your speech and speak more quietly. Or, if a client is crying, you would not want to smile; instead your nonverbals should convey caring. If you are tuned in to your client's emotions, synchronicity will happen automatically, without much conscious thought on your part. Be careful, however, not to carry this mirroring too far. For instance, if a client is very anxious, talking rapidly and loudly, you would only foster further anxiety if you were to respond in a similar fashion.

- Relax physically. You will be more open to hearing clients if your body is relaxed, and you are breathing regularly and deeply. Try to take a few minutes before a genetic counseling session to calm yourself and focus your attention.

- Use eye contact. Eye contact helps you focus on the client and indicates that you are listening. It's generally a good idea to look at your client even if s/he is not looking at you and is instead staring at the floor or looking at the wall. Eventually most clients will venture a glance at you, and when they do, it is important that they can see that you are looking at them. One disclaimer, mentioned later in this chapter, is to decrease your eye contact if it appears to be making a client uncomfortable.

- Convey sensitivity. Your nonverbals should communicate concern, alertness, and vigilance.

Problems in Attending

In our experience, problems in attending are more likely to occur when a counselor generally has either too much or too little involvement, or has an intensity that is too low or too high. Consider the following types of problematic attenders:

- The enthusiastic counselor: This counselor displays too much enthusiasm and cheeriness, such as leaning toward clients until s/he is practically in their lap, and talking loudly, quickly, and too cheerfully. At best, these behaviors can be tiring, and at worst, seem patronizing.

- The nervous counselor: This is the overly anxious counselor who avoids eye contact, fiddles with a pen or paper, or has one or more distracting mannerisms. The counselor looks as if s/he has something to hide.

- The detached counselor: This counselor is too much of a blank screen, for instance, using a piercing stare, taking a cold, clinical stance, displaying little or no emotion. The counselor comes across as analyzing the client.

- The overly concerned counselor: This counselor displays too much concern through sad facial expressions, deep sighs, etc. These behaviors may indicate that the counselor feels the clients' problems almost more than the client does. Clients may even say to this type of counselor, "Don't be so worried! I'll be OK."

- The laid-back counselor: This counselor is too casual in the genetic counseling session, slouching in the chair, yawning, dressed too casually (e.g., blue jeans). This counselor communicates that s/he is not particularly concerned about the client.

Some Cultural Variations in Attending

A nonverbal behavior can have several meanings depending on the participants and the context in which they are communicating. A nonverbal behavior in one culture may have different or even opposite meanings in another culture (Cormier and Cormier, 1991). For example, a middle-class white person from the United States might interpret lack of eye contact as a sign of anxiety or lack of interest. However, a person from China might interpret lack of eye contact as a sign of respect. Conversely, a lot of eye contact may be preferred by the U.S. client and regarded as a lack of respect by the client from China. Nevertheless, across all cultures, the best indicators of how a person feels are the way that s/he says things and how s/he looks (Van Bezooijen et al, 1983). Additionally, several studies have demonstrated that individuals from different cultures will generally agree on the types of emotions expressed in a person's face, although they will vary in their perceptions of how intensely the emotions are being felt (cf. Ekman, 1972; Markus and Kitayama, 1994).

People from different cultures have different reactions to proxemics (distance). Hill and O'Brien (1999) point out that middle class Americans regard intimate distance as 0 to 18 inches, personal distance as 17 to 48 inches, and public distance as 12 or more feet. Americans and the British tend to prefer greater distance from others than do members of many other cultural groups. For example, Latino/Latinas, Hispanics, and Middle Easterners may be more comfortable with less distance. Of course, it is important to remember that differences exist within as well as between cultures.

Another cultural difference concerns eye contact patterns. Hill and O'Brien (1999) report several cultural differences: North Americans tend to maintain eye contact when listening, and look away when speaking. Black Americans tend to look toward the person while speaking, and look away when listening. For some Native Americans and Asians, sustained eye contact is regarded as offensive and disrespectful, especially if done by younger persons to older persons. Some American Indian, Inuit, and Aboriginal Australian groups generally avoid eye contact when discussing personal topics.

You should adapt your style to your client's style; don't expect clients to adapt to yours (Hill and O'Brien, 1999). When you don't know, ask your client what feels comfortable or uncomfortable, and observe the impact of your nonverbals. For example, if you are sitting close and your client seems uncomfortable, try moving your chair back a bit and see if that helps. Also keep in mind that you should vary your attending behaviors based on both a client's immediate needs and his or her cultural background. For example, you might pull your chair closer to a client who is crying. You would slow the rate of your speech and become less animated with an extremely anxious client. You would be careful about touching clients whose backgrounds prohibit physical contact with nonfamilial members of the opposite sex (e.g., members of some Muslim sects). Some couples who seek counseling will prefer that you communicate with the husband who then translates to his wife; for instance, in some Asian and Middle-Eastern couples, the husband may prefer that you communicate through him rather than speaking directly to his wife.

Closing Comments

Although attending is the lynchpin to effective genetic counseling, it is not always an easy process. It is complicated by the fact that you must simultaneously attend to your client and to yourself.

- Attending to your clients: Remember that universally valid statements about effective physical and psychological attending and the meaning of nonverbals are impossible because of the many individual and cultural differences in clients and their situations.

- Attending to yourself: Initially you may find that focusing on behaviors that you typically do automatically will make you extremely self-conscious. You may feel somewhat awkward and mechanical as you practice attending (and other) skills. However, self-awareness is an important first step in developing your genetic counseling skills. Over time and with practice, you will learn to relax, focusing less on yourself and more on your clients.

Class Activities

Activity 1: Discussion: Think-Pair-Share Dyads

Students respond individually in writing to four questions and then discuss their responses with a partner:

1. What is attending?
2. As a genetic counselor, what attending behaviors do you think are helpful?
3. What attending behaviors are not helpful?
4. Is there a cultural group that might have an attending style different from your own?

Next, pairs of students discuss their written responses.

PROCESS

The instructor summarizes the four questions in four columns on the blackboard or on newsprint. Students then discuss their responses to each of the questions and the instructor summarizes them under the appropriate column. Then the instructor verbally summarizes major themes and presents any ideas that did not emerge from the dyads.

Estimated time: 20 to 25 minutes.

Activity 2: Attending Activity

The instructor shows photographs of people and asks students to speculate about how the people are feeling. The pictures should represent a range of emotions. Also, to make it more challenging, and to move into primary empathy, include pictures in which the person has incongruous facial expressions (e.g., an athlete who looks as if s/he is in pain after winning a sporting event).

Estimated time: 20 minutes.

Activity 3: Psychological Attending

The instructor engages in a 10- to 15-minute role-play with a volunteer from outside of the class and videotapes the role-play. The volunteer leaves and then the students recall everything they can about the client's demographic characteristics (gender, ethnicity, eye and hair color, height, weight, dress, etc.) and nonverbal behaviors. Then the instructor plays back the videotaped session and students compare it to their recalled responses. Usually the students have overlooked or forgotten some characteristics or behaviors. This leads to a discussion of how difficult it is to fully attend to all aspects of a person.

Estimated time: 30 to 35 minutes.

Activity 4: Psychological Attending

The instructor shows students a segment of a videotaped genetic counseling session (either an actual or a simulated session) with the volume turned-off. In a large group the students identify what they think the counselor and client are talking about, and what feelings each person is experiencing. Next, the instructor replays the segment with the volume on. The students compare their description of the volume-off segment with the volume-on segment.

Estimated time: 30 minutes.

INSTRUCTOR NOTE

- If students desire additional practice, they could try this same exercise at home with television programs or videotapes.

Activity 5a: Psychological Attending

Dyads discuss any topic (e.g., how it feels to be taking a course on helping skills, how it feels to be in school this year, etc.) with their eyes closed. One person is the speaker (client) and one person is the listener (counselor). They engage in a dialogue for 5 minutes. The student who is the listener (counselor) should try to sense the client's experience from what s/he says and does.

In the large group, dyads respond to these questions: What did it feel like? What cues did you respond to? What was hard? Easy? How much do you rely on visual cues? (The instructor might point out that genetic counselors who counsel over the telephone never get to see their clients).

Estimated time: 10 minutes.

Activity 5b: Psychological Attending

The same dyads take turns being the counselor and the client for two, 10-minute role-plays, using the same discussion topic as in the previous exercise. The counselor watches for a significant client nonverbal behavior (e.g., change in breathing, shift in eye contact, voice tone). The counselor should not focus on small nonverbal behaviors that are out of context with spoken words. After the counselor has observed one significant nonverbal behavior, s/he should ask the client if the client is aware of what is happening to his/her breathing, etc. The counselor *should not interpret* or assign meaning to the client's behavior. The counselor should merely notice where the focus takes the client.

After both role-plays are completed, the students discuss the following questions: What kinds of nonverbal behaviors did you focus on? What happened when you pointed them out? Both the counselor and the client should comment on the latter question.

Estimated time: 30 minutes.

Activity 6: Physical Attending Role-Play

Dyads engage in 10- to 15-minute role-plays, taking turns as counselor and client. The instructor gives the client the following instructions (the counselors should not see these instructions).

Role-play 1: The client violates the counselor's personal space (sits too close, leans in too far, touches the counselor).

Role-play 2: The client acts uncomfortable with the counselor's eye contact. Whenever the counselor looks at the client, the client should turn away, fidget, stammer, etc.

After both role-plays are completed, the students discuss the following questions: How did the counselor feel? What did the counselor in each role-play think was going on? Was the counselor in role-play 2 aware of the client's discomfort with eye contact? What did the counselor do with this awareness?

Estimated time: 30 to 40 minutes.

Activity 7a: Low-Level Attending Skills Model

The instructor and a volunteer genetic counseling client engage in a role-play in which the counselor demonstrates poor physical and psychological attending behaviors (see Problems in Attending, above, for some examples of poor attending). Students observe and take notes of examples of poor attending.

Estimated time: 5 minutes.

PROCESS

Students discuss their examples of poor attending and the impact of the counselor's poor attending skills on the client. The client can offer her or his impressions of the counselor's behaviors after the other students have made their comments.

Estimated time: 10 minutes.

Activity 7b: High-Level Attending Skills Model

The instructor and the same volunteer repeat the same role-play, only this time the counselor displays good attending skills. Students take notes of examples of good attending behaviors.

Estimated time: 5 minutes.

PROCESS

Students discuss their examples of good attending and the impact of the counselor's attending behaviors on the client. They also contrast this role-play to the low-level role-play.

Estimated time: 10 minutes.

INSTRUCTOR NOTE

• Students could work together in Think-Pair-Share dyads to identify attending examples and their impact on the client before discussing them as a whole class.

Activity 8: Physical Attending Triad Exercise

Three students practice physical attending skills in 5-minute role-plays taking turns as counselor, client, and observer. Allow 10 minutes of feedback after each role-play. The students should focus on using good physical attending behaviors. Students can refer to the section on Effective Counselor Physical Attending Behaviors for examples of good skills.

PROCESS

In the large group discuss the following: How was it to do this exercise? What are you learning about attending in general? About yourself?

Estimated time: 60 to 75 minutes.

INSTRUCTOR NOTE

- Some individuals have very powerful nonverbals (e.g., a strong, loud voice, piercing gaze, etc.). If this is the case for any of the students, the instructor should give them feedback about how they might tone down these intense nonverbals.

Written Exercises

Exercise 1*

Briefly describe two possible meanings for each of the following client nonverbal behaviors:

Client Nonverbal Behavior   Possible Meaning

- Client stares at the floor.
- Client grimaces at the term "defect".
- Client jiggles her foot repeatedly.
- Client sighs deeply and says nothing.
- Client has drops of sweat on his forehead.
- Client grips her partner's hand.
- Client leans away from the counselor.
- Client stumbles over his words.
- Client frowns.
- Client draws in a deep breath.

Exercise 2

Respond in writing to the following questions:

- What physical attending behaviors are most difficult for you to use?

---

*Adapted from Cormier and Cormier, 1991.

- What physical attending behaviors are the easiest for you to use?
- What are your most powerful nonverbal behaviors (i.e., what do other people notice the most about your nonverbal behaviors)?
- What distracting nonverbal behaviors do you engage in when you feel nervous? Bored? Distracted?

INSTRUCTOR NOTE

- Students could write responses to the four questions as part of a journal or a case log book.

Annotated Bibliography

Bachelor, A. (1988). Clients' perceptions of the therapeutic alliance: a qualitative analysis. Journal of Counseling Psychology, 42, 323-327.
[Discusses the impact of nonverbal behaviors on the strength of the helping relationship.]
Baker, D. L. (1998). Interviewing techniques. In: Baker, D. L., Schuette, J. L., Uhlmann, W. R., eds. A guide to genetic counseling. New York: John Wiley & Sons, 55-74.
[Describes genetic counselor effective attending.]
Cormier, W. H., Cormier, L. S. (1991). Interviewing strategies for helpers: fundamental skills and cognitive behavioral interventions (3rd ed.). Pacific Grove, CA: Brooks/Cole.
[Provides a detailed description of attending skills in the helping professions and contains several exercises for practicing attending skills.]
Egan, G. (1994). The skilled helper (5th ed.). Monterey, CA: Brooks/Cole.
[Contains a detailed discussion of psychological and physical attending and provides skills practice exercises.]
Ekman, P., Friesen, W. V. (1984). Unmasking the face (reprint ed.). Palo Alto, CA: Consulting Psychologists Press.
[Reports the results of research describing how emotions are expressed in the face.]
Fine, S. F., Glasser, P. H. (1996). First interview: Establishing an effective helping relationship. Thousand Oaks, CA: Sage Publications.
[Provides specific tips and suggestions for effective attending.]
LaFrance, M., Mayo, C. (1976). Racial differences in gaze behavior during conversation: two systematic observational studies. Journal of Personality and Social Psychology, 33, 547-552.
[Reports research findings suggesting differences in eye contact as a function of racial background.]

CHAPTER 4

Listening to Clients:
Primary Empathy Skills

Learning Objectives

1. Define primary empathy and its functions in genetic counseling.
2. Distinguish between different types of primary empathy responses.
3. Develop primary empathy skills through practice and feedback.

Definition of Empathy

Empathy consists of understanding what another person is experiencing and communicating that understanding to the person (Egan, 1994). It is the capacity to put yourself in another person's place to understand from her or his frame of reference (Bellet and Maloney, 1991).

Although there are many different definitions of empathy, most emphasize two major dimensions: empathic emotions (having an emotional reaction that is in tune with the client's experience) and intellectual empathy (engaging in role-taking or perspective-taking) (Bellet and Maloney, 1991; Duan and Hill, 1996; Gladstein, 1983). Empathy forms the very basis of all human interactions (Duan and Hill, 1996), and it is an essential condition within Carl Rogers' person-centered approach to counseling.

Most authors (e.g., Barrett-Lennard, 1981; Duan and Hill, 1996; Gladstein, 1983) agree that empathy is a multistage, interpersonal process that contains at least three elements:

- Can I sense what you experience?
- Can I communicate this sense to you?
- Can you perceive this communication as my understanding you/your experience? This third component depends to a great extent on the characteristics of the recipient of the empathy response (Barrett-Lennard, 1981).

Types of Empathy

There are two major types of empathy responses: primary empathy and advanced empathy. Primary empathy communicates initial understanding of what a client is experiencing. Primary empathy is particularly important for rapport building and problem exploration. In genetic counseling, you use your own words to convey an understanding of surface, fairly explicit client experiences.

Example: Client (tearfully says): "I don't want anything to be wrong with my baby."

Counselor: "It's upsetting to think that something could be wrong."

Advanced empathy communicates an understanding of underlying, implicit aspects of client experience. This type of empathy is useful for dynamic understanding (assessing the client's deeper, less obvious feelings and experiences). Your responses are additive in that they go beyond surface client expressions.

Example: A client is focusing on being afraid to have an amniocentesis because she dislikes needles. You believe there is more to her reaction and say, "In addition to being scared about the amnio procedure, maybe you're also frightened about what the test results might reveal."

This chapter discusses primary empathy. Chapter 8 considers advanced empathy in more detail.

Empathy Is Not Feeling What the Client Feels

An important aspect of empathy is the ability to maintain some distance from your client's experience. Empathy involves sensing "the client's world as if it were your own, but without ever losing the 'as if' quality" and sensing "the client's anger, fear, or confusion as if it were your own, yet without your own anger, fear, or confusion getting bound up in it" (Rogers, 1992, p. 829). If you lose the *as if* quality, then the result is not empathy but identification with the client (Barrett-Lennard, 1981), or what some authors (e.g., Kessler, 1998) refer to as countertransference (see Chapter 12). You no longer understand your client's experience, but rather you become caught up in your own feelings and perceptions. Another problem with counselor identification occurs when your emotional reaction to the client is too intense, what Gladstein (1983) calls *empathic distress*. In this situation, you are likely to psychologically *move away* from the client (e.g., avoiding a discussion of client feelings, offering false reassurance that everything will be OK, etc.).

Functions of Primary Empathy

Primary empathy serves several functions:

• Reinforcing the client to continue talking.

- Providing clarification for both the counselor and the client.
- Making the counselor seem similar to the client, thus increasing social attractiveness (the counselor is viewed as warm and likable).
- Providing a model for the client of how to be empathic.
- Facilitating the establishment of rapport and trust.
- Helping clients feel understood by the counselor.
- Helping clients manage their feelings.
- Facilitating client risk-taking; discussion of unpleasant emotions; and reduction of nonproductive anger, overwhelming anxiety, or other strong feelings (Cormier and Cormier, 1991; Danish et al, 1980; Egan, 1994; Hill and O'Brien, 1999).

Empathy Is Both Innate and Learned

Many theorists and researchers believe that the essence of one's ability to sense another person's experience is present in infancy and develops throughout childhood and adolescence (e.g., Bryan, 1972; Gladstein, 1983; Selman, 1980). For instance, have you ever noticed that in a nursery or day-care setting, when one infant becomes distressed and cries, other infants begin to cry as well? Some researchers believe that this behavior is evidence of the beginnings of empathy (Azar, 1997). As children mature, through their interactions with parents and others, they may be socialized to resonate with other people's feelings and perspectives. Once a sufficient level of cognitive ability develops, the older child/adolescent is capable of communicating this understanding in a more sophisticated manner. Although the ability to understand another person's experience probably cannot be taught to adults, adults can, through appropriate learning activities, develop better skills for communicating their understanding to others (Gladstein, 1983). They also can learn to focus their empathic ability.

Tapping Into Empathic Understanding

There are several things that you might do to understand your clients:

- Put yourself in your client's place and ask yourself how you might think and feel. This is called *perspective taking*.
- Relate your client's experience to a similar life experience of your own.
- Pay attention to client verbalizations. Listen to what your client is saying.
- Attend to client nonverbal behaviors. As we previously discussed, nonverbals may account for as much as 93% of the message.
- Become aware of your own nonverbal reactions as you listen to a client. Is your stomach tightening? Do you find yourself close to tears? Are you sighing?

- Gain experience. As you see more clients, you will recognize that certain experiences and emotions frequently coincide with particular genetic and psychosocial conditions.

- Read the genetic counseling literature. You can gain at least an intellectual understanding of what it might be like to be a genetic counseling client by reading about different genetic conditions, family challenges, and client experiences with genetic counseling and decision making.

- Draw upon your intuition. Play a hunch: if you suspect that your client is feeling or thinking something, tentatively suggest it.

Effectively Communicating Empathic Understanding

- Acknowledge and set aside your biases. It's virtually impossible to communicate authentic empathy when you feel judgmental.

- Feel comfortable with your client. If you feel threatened or defensive, then you will have difficulty experiencing and expressing empathy (Barrett-Lennard, 1981).

- Focus on your client rather than on yourself. With practice and experience you will gradually become less self-conscious and more client-focused. If you find yourself becoming overly self-conscious, try taking a couple of deep breaths, and relax in your chair. You will be in the flow when you are no longer thinking, "How do I sound? What am I going to say next? What do I do if my client cries?"

- State your understanding of your client's experience concisely and in your own words. You should highlight the essence of what your client has expressed, rather than giving a verbatim account. For example: The client says, "I just found out that my sister's baby has this genetic condition. I can't believe that my baby is at risk. I'm afraid of what the prenatal tests are going to show. If only I'd known this before I became pregnant!" You might say: "You're very worried about your baby." This response emphasizes what you believe to be the most salient aspect of her concerns.

- State your empathy tentatively. Tentative statements give the client room to correct you if you are off-target. For example: "Is it possible that. . .?"; "Maybe you are feeling...?"; "Perhaps you feel...?"; "So it sounds as though you might be feeling...?").

- Aim for the ballpark rather than the bull's-eye. Although it's always gratifying when a client responds to your statement by saying, "That's exactly it," it's sufficient if you have approximately the same intensity and are using words that are close in meaning to your client's. A ballpark goal reduces some of the pressure on you and frees you to focus on your client.

- Communicate acceptance. Fine and Glasser (1996) wisely point out that, "Feelings belong to the person who has them and are neither right nor wrong. . . .Feelings are useful, even negative and painful feelings. They cannot and should not be argued or debated away" (p. 60). They further state that communicating acceptance means "not debating, not arguing, never using the b word: but" (p. 60).

- Reflect content and affect. Aim for responses that tap both the content (intellectual empathy) and feeling (emotional empathy) dimensions of your client's experience (e.g., "It sounds like you're angry because your parents never told you about your risk for this condition").
- Be thorough. State all sides of your client's message, including conflicting and contradictory parts; for some situations where your client is mixed or conflicted, try saying, "You really feel both ways, don't you?" (Martin, 2000).
- Respond empathically and then stop! Clients will almost always respond. Beginning counselors frequently make the mistake of giving a great empathic response and then immediately following it with a question, not allowing the client an opportunity to respond.

Primary Empathy Responses

Primary empathy responses vary on a continuum from head nods, silence, and minimal encouragers, to more complex responses that reflect the content and feeling of client experiences.

The Primary Empathy Continuum

Simple						Complex
Silence	Minimal Encourager	Paraphrase	Summary	Reflect Content	Reflect Feeling	Content and Affect Reflection

Pedersen and Ivey (1993) identify six major types of primary empathy responses:

Minimal Encouragers

Minimal encouragers prompt clients to continue talking, but do not interrupt the flow of the session. They may include head nods, hand gestures, brief comments such as "Uh-huh" and "Mm-hm," repeating a few key words, and even silence. A minimal encourager response is "the simplest of the listening skills—but it is also one of the most powerful. Research indicates that effective and experienced counselors use this skill significantly more often than ineffective and inexperienced helpers" (Pedersen and Ivey, 1993, p. 121).

Example: Client: "It's my first baby and I'm afraid something's going to go wrong."

Counselor: "So, you're afraid?"

Paraphrasing

Paraphrases give clients feedback on the essence of what they said. To paraphrase, you use your own concise words, while including some of the client's key phrases (Pedersen and Ivey, 1993).

Example: Client: "We want to adopt a child, but we're worried about what limitations this genetic condition will place on the child we've been offered."

Counselor: "You want to know what you'd be taking on if you adopted this child?"

Summarizing

Summaries are somewhat longer responses than paraphrases, and they contain more information (Pedersen and Ivey, 1993). When clients have disclosed a large amount of information, summarizing is useful because it synthesizes disparate parts of their stories. Furthermore, "Many cultures use story-telling as a major vehicle in disseminating their culture" (Pedersen and Ivey, 1993, p. 119). When working with a client who uses a story telling vehicle, you may be able to communicate your understanding with a summary that ties together the major themes of the story.

Example: Client: "I really want to be tested for this gene. My mom and her sister died of breast cancer, and my sister was diagnosed two months ago. I have two daughters myself. What will happen to them?"

Counselor: "So, what I'm hearing is, you're very scared. And with all of the cancer in your family, you're worried about getting it and passing it on."

Content Reflection (Content Responses)

Content responses emphasize the cognitive aspects of a client's experience. Content responses help clients explore their goals and values and gain deeper understanding of their experience. They also allow counselors to identify "different or conflicting culturally learned perspectives without necessarily resolving them in favor of either viewpoint" (Pedersen and Ivey, 1993, p. 156).

Example: A deaf couple wants a nonhearing child. However, their desire may conflict with the values of the broader culture.

Client: "We don't know how to raise a child to live in the hearing world. We're fully prepared to raise a deaf child. But you can't believe what people are saying about our decision!"

Counselor: "You're kind of caught between your deaf culture and the judgments of others."

Feeling Reflection (Affective Responses)

Affective responses stress the emotional aspects of a client's experience.

Example: Client: "I've tried to find out information about Trisomy 18, and
nobody's telling me what I need to know!"

Counselor: "You sound frustrated and angry."

Content and Feeling Reflections

These responses combine statements about client feelings and the situations/
factors contributing to these feelings. One strategy for making this kind of
reflection is to say, "You feel...because..." (Egan, 1994).

Example: Client: "I'm glad we went ahead with the prenatal testing. I can't wait to
tell my husband the good news! We were so worried about our baby
having hemophilia."

Counselor: "You feel really relieved because the results are negative."

The Importance of Attending to Client Affect

Feelings are a universal human experience. Izard (1977) identified 10 universal
emotions: interest, enjoyment, surprise, distress, anger, disgust, contempt, fear,
shame, and guilt. These feelings can be broadly categorized by domain and by
intensity:

- Domain: The type of emotion. Feelings are either pleasant (e.g., happy) or
  unpleasant (e.g., sad).
- Intensity: The degree or level of emotion. Feelings can range on a continuum
  from mild (e.g., irritated) to moderate (e.g., angry) to extreme (e.g., furious).

Although feelings may be experienced universally, the ways in which clients
express their emotions are very culture-specific. Also, the events and situations
that trigger specific emotions have a strong cultural component.

Client feelings are a critical part of the genetic counseling process (Kessler,
1999). You should not "shy away from interviewing in a manner that may allow
the client to access the emotional dimension of her experiences" (Baker, 1998,
p. 72). Some clients may use feeling words frequently but not *hear* them. When
reflected back to them, they may be able to hear their feelings and come to
terms with them. Furthermore, clients often "have mixed feelings and are not
sure themselves about what they feel. This confusion can interfere with an
interview unless the feelings are clarified." (Pedersen and Ivey, 1993, p. 150).
Primary empathy allows you to recognize and label client feelings. When you
reflect client feelings (in a tentative way), you help to clarify them. However, be
sure to observe client reactions, as they may indicate the accuracy of your
reflections. To recognize and label feelings accurately, Pedersen and Ivey (1993)
suggest the following:

- Listen for feeling words.
- Watch for clues to feelings in client nonverbals.

- Reflect feelings back to the client in your own words.
- Use a basic sentence stem that reflects the way the client talks (e.g., "It looks like. . ." for a visual client; "It sounds like. . ." for an auditory client; "It feels like. . ." for a tactile client).
- State the situation in which the feelings occurred (e.g., "It looks like you feel. . .because. . .).
- Check for accuracy.

Cultural Empathy

Given the growing attention to and awareness of cultural differences, some authors are beginning to further refine the concept of empathy. Ridley (1995; Ridley and Lingle, 1996) coined the term "cultural empathy" to describe empathy that is sensitive to client culture. Cultural empathy is based on three principles:

- Every client should be understood from her or his unique frame of reference.
- Normative information (e.g., data about what people do on average, statistics about the typical person) does not always fit a particular client, although it can be useful as background information.
- People are a dynamic blend of multiple roles and identities.

Cultural empathy involves two components: understanding and responsiveness. Cultural empathic understanding entails:

1. Cultural self-other differentiation: examining your own cultural identity and values and learning about the client's cultural identity and values.
2. Perspective taking: developing an understanding of the client's culture, and using this understanding to see the client in his/her cultural context.
3. Probing for insight: collecting information about the client's entire self-experience and asking clarifying questions. (Open-ended questions, discussed in Chapter 5, can be useful.)

    Cultural empathic responsiveness entails:

1. Conveying interest in learning more about the client's cultural values (e.g., "Tell me more about how this is viewed in your culture").
2. Expressing naivete with respect to the client's cultural experience (e.g., "What is it like for you as a member of this religious community?").
3. Verbally disclosing to the client your understanding of the client's self-experience (e.g., through paraphrasing, reflecting, summarizing).

Pedersen (1991) has identified several skills for effectively empathizing with clients whose cultural backgrounds differ from your own. One of these skills involves recognizing positive aspects in what you personally may perceive as a negative experience (e.g., a Triple Screen indicates a high risk of having a child with Down syndrome; the Catholic Hispanic couple wants a child under any condition. They do not perceive a high risk of Down syndrome as negatively as you might). A second skill is recognizing potential negative implications from what you personally consider to be a positive experience (e.g., a couple who both have achondroplasia have a child who is not affected and will therefore be of normal height. The couple may react with mixed emotions). Cultural empathy requires you to see the world the way clients see it, rather than imposing your own take on the situation. Culturally empathic counselors are able to step out of their own frame of reference to view a situation the way the client views it. They seek information about their clients' beliefs by paying careful attention to clues about their values (Brown, 1997).

Common Empathy Mistakes

Empathy mistakes tend to be due to either covert processes (beliefs, assumptions, attitudes, etc.) or to overt processes (actual behaviors).

Mistakes Due to Covert Processes

- Overidentifying: When you overidentify, you feel too much with your client. You must be able to step back after sensing a client's feelings, thus giving yourself some distance from the affect (Barrett-Lennard, 1981; Rogers, 1992). It's usually ineffective if you feel as angry or sad as clients do about their situations. One way to understand and address overidentification is to discuss it with your clinical supervisor.

- Making assumptions: You should be careful about assuming that your clients will feel exactly what you would feel, since this may not be the case. For example, you might feel devastated if your baby had Down syndrome. However, a couple who desperately wants to have a child may feel very disappointed, but also happy that they will be parents.

- Being afraid of client feelings: During genetic counseling, clients may experience a range of emotions. Indeed, McCarthy Veach and her colleagues (1999) found that all of the 28 former clients in their study of the impact of genetic counseling experienced strong emotions during their genetic counseling sessions. Almost two thirds of the feelings were unpleasant (e.g., anxiety, anger or frustration, confusion, sadness and disappointment, discomfort, shame and guilt), while about one third were positive (e.g., comfortable, relief, felt supported). Sometimes our fear comes from mistaken beliefs. For instance, you (and your clients) may think that if certain feelings are identified (e.g., grief, anger), you will both be overwhelmed by them; or you might believe that only some feelings are acceptable (e.g., sadness), while others are not (e.g., anger, despair). In addition, you may mistakenly believe that you must fix your client's feelings, or that if they are expressed, you won't know how to handle these emotions. It is important to remember that there are no right or wrong feelings. People feel what they feel. Furthermore, when clients express emotions, they usually end up feeling more in control. Until clients can get out their sadness, anxiety, anger, etc., they will be less able to digest the information that you provide and less capable of making decisions.

- Thinking you can't understand if you haven't had your client's experience: Although you may never have had a family member with a genetic condition, you do have experiences in your own life of loss, disappointment, grief, etc. Your own experience will help you empathize with client feelings. Of course you may lack a specific understanding of what it's like to be in a particular client's situation, for example, what it's like to live with a parent who has Huntington disease. In this case you might say, "Please tell me about what this is like for you so that I can try to understand."

- Thinking that clients are different from nonclients: They aren't. Clients have the same hopes, fears, anxieties, and beliefs as anyone. Although they might differ from most other people because they have come for assistance with a possible or known genetic condition, the empathic process is the same for genetic counseling clients as it is for all people. As we stated earlier, empathy is an essential ingredient in all human relationships (Duan and Hill, 1996). Ask yourself, if you were sitting with a friend who thought s/he might have a genetic condition, what would you say to empathize? Try using a similar response with your clients.

- Assuming that all clients will respond in the same way: It is important to remember that people may have different reactions to similar events. One client may be angry, another sad, etc. Clients will also respond in unique ways to your interventions. Therefore, you cannot give a cookbook response and get positive results with all of your clients. There is no answer key in genetic counseling, because clients differ in subtle and not-so-subtle ways. For example, consider a prenatal client who has four children and discovers that her current pregnancy has Trisomy 18. How might her experience be similar to and different from the prenatal client who has undergone several years of infertility treatment, has no children, and discovers that her pregnancy has Trisomy 18? Or consider the case of a couple from China who will be returning to China. They have one daughter, and their second baby is a boy with a fatal disease. In Chinese culture male children are valued highly. The hopes of Chinese parents rest in having a son to support them in their old age and to carry on the family name. Consider how this couple's experience would be different if they were white and from the U.S.

- Assuming that all clients desire the same type and amount of primary empathy: Clients vary in their preferences for counselor empathy (Duan and Hill, 1996; Gladstein, 1983). Some clients, wanting a close relationship, will desire more affective empathy from you, while others who wish for a more neutral emotional relationship will prefer less affective empathy (Gladstein, 1983).

One clue about the type of relationship that a client wants is his/her reactions to your initial affective empathy statements. If clients do not elaborate and/ or seem nonverbally taken aback, these may be indications that they do not wish you to make a lot of overt empathy expressions. You should not take this personally or as a sign that something went wrong. Not every client wants the same type of relationship, and not every client will discuss feelings no matter how many techniques you pull out of your bag.

Mistakes Due to Overt Processes

- Not replying: Suggesting that what your client said is not worth a response or is not relevant to genetic counseling (Weil, 2000). It is a matter of common decency to display some understanding of your client (Kessler, 1999).

- Using cliches: "You can always try to have another baby"; "Time heals all wounds"; "New treatments are coming along all the time."

- Focusing too much on either content or affect: Beginning counselors tend to emphasize content and overlook affect. In addition, Western cultures tend to stress intellect, often at the expense of feelings. To further complicate matters, clients might avoid expressing feelings because they are afraid of losing control and/or are not sure that discussing feelings is appropriate in genetic counseling (McCarthy Veach et al, 1999). On the other hand, sometimes counselors emphasize clients' feelings at the expense of content. Too much attention to feelings can prevent you and your client from moving to goal setting and decision making. Effective genetic counseling includes a balance of attention to content and affect.

- Prematurely using advanced empathy: Even if your remark is on target, it may be too threatening unless you have established an initial rapport with your client, and your client is ready to hear your interpretation.

- Inaccurate labeling/distorting: Making statements that are wrong or miss the mark with respect to either the feelings or content of your client's experience (e.g., reflecting to a client that she seems to know what she wants to do when she has repeatedly stated that she can't decide or telling a furious client that he seems a bit irritated).

- Pretending to understand: This is not genuine, and clients will pick up on this.

- Parroting: Primary empathy is not simply repeating client words verbatim. You should communicate the core or essence of client expressions and do so in your own words.

- Jumping in too quickly: You should avoid interrupting clients or thinking that you must respond immediately after they speak.

- Using inappropriate language: Jargon or words that are too complicated can distance clients. Try to match your client's language level.

- Being longwinded: Long, rambling primary empathy statements often confuse clients. Remember to keep your responses concise and to the point.

- Forgetting to use silence: Silence and a calm presence can be as empathic as anything you might say. Silence allows clients to become aware of their feelings and to digest what has been discussed. It also helps them to collect their thoughts.

- Not allowing enough time for clients to respond to your empathy statements: Silence often seems much longer to a beginning counselor than it really is. Try mentally counting to 10 or 15 before jumping in; by then, most clients will have responded to your statement. Of course when clients cannot think of how to say what they want to express (watch for nonverbal indications of this), you can step in with a statement like, "It's hard to put into words what you want to say"; or "What do you think about what I just said?"

- Using empathy responses inappropriately: Reflections can encourage clients to continue talking. So when you are trying to change topics, want a client to stop talking, or wish to end the session, you should generally avoid reflections.

- Taking sides: Expressing a lot of empathy for one member of a couple or for certain family members and forgetting to do so with the other individuals is divisive and judgmental on your part.

- Forgetting the cultural context: Empathy can be conveyed quite differently for some populations (e.g., studies have shown that some cultural groups, such as Southeast Asian immigrants, tend to manifest depression and anxiety through somatic complaints rather than by using feeling words). Some clients may prefer to project their problems onto less personal or more external sources, so you need to proceed more slowly and indirectly in exploring their situation (Ishiyama, 1995). For instance, you might say, "Some clients might feel. . ." Salzman (1995) suggests that counselors can "improve sensitivity to and understanding of their clients if they constantly ask themselves how the person they are working with is like all other humans (the universal aspect) and like no other human being (the idiosyncratic aspect)" (p. 190). One must recognize the validity of cultural variation, overarching human commonalities, and within-group variation (Salzman, 1995).

- Making a content response when you intended to reflect feelings: If you wish to make an affective reflection, then be sure you identify a feeling. If your response begins, "You feel that. . .," it is probably reflecting the client's thoughts and not feelings.

Typical Concerns About Primary Empathy

Beginning counselors typically have a number of questions about empathy. In the following sections, we address some of these concerns.

Why Is Empathy Sometimes Difficult?

Many clients experience intense, unpleasant emotions. It is a natural human tendency to want to feel pleasant emotions and to want to avoid unpleasant ones. So clients may avoid expressing and/or acknowledging their unpleasant feelings. Similarly, we may avoid empathizing because we are afraid of feeling our clients' pain.

Reflecting feelings is challenging because, "In our culture [U.S.] we learn to comfort people by encouraging them to run away from their feelings. We are taught to say, 'Don't cry, it'll be all right,' when it quite probably won't be all right and the person really needs to cry to release his emotional pain" (Geldard, 1989, p. 37). Although directly confronting feelings may be uncomfortable for your clients, it is sometimes essential because, "At crisis times in our lives the emotional pressure builds up until we are ready to explode. In this state our thought processes are blocked and we are unable to cope. We feel out of control of ourselves. To regain control, we must first release some of the emotional pressure" (Geldard, 1989, p. 39).

Your own background, abilities, and present situation affect the ease with which you can empathize. Your empathy toward a client is impacted by your "empathic capacity, past experience, motivation to empathize, and affective and cognitive state at the time of the session" (Duan and Hill, 1996, p. 268).

Is Empathy Different from Sympathy?

Empathy is not the same thing as sympathy. Sympathy is *feeling sorry for* a person and it conveys a one-up/one-down relationship. Empathy is *feeling with*, and it conveys a more equal, collaborative relationship. Sympathy implies pity, while empathy implies trust: "I am saying to my client that he or she is strong enough to solve problems, that I will not condescend pityingly, and that our work together is not just hand holding" (Martin, 2000, p. 9). Also, sympathy can be problematic because the sympathetic counselor may feel so emotionally uncomfortable that s/he will try to remove this discomfort and *forget* about helping the client (Allport, 1924).

Is Empathy Affected If I Have an Experience Similar to My Client's?

There can be both advantages and disadvantages to having a similar experience to your client. It can be an advantage to have some first-hand experience of what it might be like to be this person. Also, you may gain credibility with a client who is more likely to think that you can really understand. It can be a disadvantage if you impose your experience on a client, rather than listening to how it is for her or him. Additionally, the client may not disclose as much information because s/he assumes that you know it by virtue of having had a similar experience. As we said earlier, you can have empathy without having a similar experience. There is no need to feel inadequate or apologetic for not having had a similar experience. If a client asks you whether you've had a similar experience, you could try matter-of-factly responding, "No, I haven't. I have worked with clients who have had similar situations, but I'd like to try to understand how it is for you." (Of course, this response only works if you've actually had clients!) Or, try saying, "No, I haven't, but I'd really like to understand what it's like for you. Would you please tell me about it?"

Won't My Clients Think I'm Just Parroting Their Words?

No, as long as you are using concise responses that are expressed in your own words. While your responses may seem forced or awkward to you, they allow clients to hear (sometimes for the first time) what they are experiencing and what they sound like.

What Can I Accomplish with Reflections?

Sometimes students tell us that they feel as if they aren't accomplishing anything if they only repeat what the client says. Their comments reflect a Western perspective in which problem solving is highly valued. What they don't realize is that empathy is a powerful component of effective problem solving. It provides an initial understanding of the nature of client concerns and helps clients gain insight about their attitudes, feelings, and real issues. Without accurate empathy, you may come up with a great solution. However, it may be a solution to the wrong problem (Egan, 1994)!

Additionally, we agree with Kessler (1999), who stated:

The new genetics increasingly confronts professionals with issues that tend to place them into the role of counselors or therapists rather than educators. The fact that the information they provide is emotionally evocative and intimately connected to the survival of clients requires counselors to make a deeper exploration of the personal meaning clients give to the information. [p. 341].

Empathy is an essential skill for this type of exploration.

Is There Anything I Should Avoid Saying If I Want to Be Empathic?

We recommend that you never say any of the following to your clients:

"I know exactly how you feel." You don't. You can never get inside another person's skin and have the exact experience. This type of remark can be quite insulting to clients. Furthermore, if clients actually believe it, then they may stop describing their experience because they assume that you already know all about it.

"You shouldn't feel that way." This statement is judgmental and may make clients feel that they are wrong to feel the way that they do. Furthermore, the reality is that the client *does* feel that way.

"Everybody feels (or would feel) that way." A remark like this can trivialize the client's experience.

"Nobody feels (or would feel) that way." This is another judgmental response suggesting that clients are wrong to feel the way that they do.

"Why do you feel that way?" "Why" implies judgment and suggests that the client's feeling is not appropriate.

Closing Comments

Rogers (1992) said that empathy and other essential therapist qualities are "qualities of experience, not intellectual information. If they are to be acquired, they must, in my opinion, be acquired through experiential training" (p. 831). We agree with Rogers that empathy will come as you gain experience—lots of it! As you see more genetic counseling clients, you will grow in your capacity to understand them. Also, you may be able to speed up the process by experimenting, trying different approaches, and discussing the outcomes with your clinical supervisors. You will begin to appreciate the diverse responses of clients. You will also recognize subtle differences in emotional responses due to individual circumstances, and counsel each client according to his or her specific situation.

Class Activities

Activity 1: Think-Pair-Share Dyads

Students respond individually in writing to three questions:

- What is empathy?
- Where does empathy come from?
- What functions does it serve in genetic counseling?

Next, pairs of students discuss their written responses.

Estimated time: 10 to 15 minutes.

PROCESS

Dyads report on their discussion. Instructor summarizes major themes and presents any ideas that did not emerge from the dyads.

Estimated time: 10 to 15 minutes.

INSTRUCTOR NOTE

- This activity could also be done in groups of three or four students. Or it could be assigned as a paper. Instead of writing individually, students can think about their responses to the questions for a couple of minutes and then discuss with a partner.

Activity 2a: Low-Level Empathy Skills Model

Instructor and a volunteer genetic counseling client engage in a role-play in which the counselor demonstrates poor primary empathy (refer to Common

Mistakes in Primary Empathy, above, for ideas on how to model a low skill level). Students observe and take notes of examples of poor empathy.

Estimated time: 5 minutes.

PROCESS

Students discuss their examples of poor empathy. Then they discuss with the instructor the impact of the counselor's poor empathy on the client. The client can offer her or his impressions of the counselor's behaviors after the other students have made their comments.

Estimated time: 10 minutes.

Activity 2b: High-Level Empathy Skills Model

Instructor and the same volunteer repeat the same role-play, only this time the counselor displays effective primary empathy skills. Students take notes of examples of good empathy skills.

Estimated time: 5 minutes.

PROCESS

Students discuss their examples of good empathy. Then they discuss with the instructor the impact of the counselor's empathy on the client. As part of the discussion they contrast this role-play to the low-level role-play.

Estimated time: 10 minutes.

INSTRUCTOR NOTE

- Students could work together in Think-Pair-Share dyads to identify empathy examples and their impact on the client.
- Students can also comment on attending skills, but the focus of the processing should be on primary empathy.

Activity 3: Brainstorming Exercise

Small groups of students brainstorm cliches that might be said to genetic counseling clients (e.g., "Time heals all wounds"). This exercise could also be done in dyads or individually. Students discuss their cliches with the large group and then discuss with the instructor why people use cliches and what impact they might have on genetic counseling clients.

Estimated time: 15 to 20 minutes.

INSTRUCTOR NOTE

- A variation on this activity is to interview genetic counselors and ask them to describe the most offensive and outrageous things other people have said to their clients about their genetic situations.

Activity 4: Domain/Intensity Exercise

As stated in the chapter, in order to be effective, an empathic response needs to be accurate with respect to domain (positive or negative) and intensity (level of emotion). Students write a four- or five-sentence description of a concern they might have if they were a genetic counseling client. They should not state any feelings about the concern in their description. Next the instructor asks for a student volunteer to read her/his sentences to the class matter of factly without conveying feelings either verbally or nonverbally. Students brainstorm possible feelings. As each feeling is identified, the instructor asks whether it is positive or negative, and mild, moderate, or intense. The instructor lists students' feeling words on a blackboard or newsprint as follows:

	Positive Feeling Domain	Negative Feeling Domain
Mild Intensity		
Moderate Intensity		
Strong Intensity		

After the students have finished brainstorming, the instructor ask the volunteer to select the feelings that come closest to what s/he would feel as this genetic counseling client.

This activity can be repeated with several volunteers.

PROCESS

Students respond to the following questions:

- What was challenging about this activity?
- Were you able to choose the right domain? The right intensity?
- Can you see how easy it is to choose the wrong intensity?
- Do you think it's more ineffective to choose the wrong intensity or the wrong domain?

Activity 5: Round Robin Exercise

ROUND 1

Have students work in groups of five or six. The instructor goes first to demonstrate the activity. The instructor asks the student beside her/him, student A, to make two or three statements that a genetic counseling client might make about her/his situation. Then the instructor mirrors the client statement by repeating exactly what the client said. Then student A turns to student B, who makes a different two or three sentence client statement, and student A mirrors student B's client statement. Continue this process around the circle until everyone has had the chance to be both counselor and client.

Round 2

For the next round, using the same client statements, the counselor makes a concise, own-words response summarizing the most important content without labeling feelings.

Round 3

For the final round, using the same client statements, the counselor makes a content and affective response. The counselor should use the formula, "You feel. . .because. . ."

Estimated time: 20 to 30 minutes.

Instructor Note

• The instructor should remember to reinforce good responses. If a student's response is way off base, ask the group, "Does anyone else have some other ideas about how to respond?

Activity 6: Small Group Role-Play

Form small groups (four to five students). In each group, either the instructor or a student pretends to be a genetic counseling client. The client begins by talking about why s/he came for genetic counseling. Next, one of the group members, student A, counsels the client for two or three interchanges in which student A uses primary empathy responses. Then the next student, student B, becomes the counselor and counsels the client for two or three responses. The process continues until each student has had the opportunity to be counselor. Halfway through the process, the group should stop and have a discussion about what they know about the client so far, what else they may need to cover via primary empathy, etc. This discussion can also be done whenever a student gets stuck while in the counselor role.

Estimated time: 30 to 40 minutes.

Activity 7: Triad Exercise

Three students practice primary empathy and attending skills in 15-minute role-plays taking turns as counselor, client, and observer. They spend 10 minutes for feedback after each role-play. The observer should stop the counselor if the counselor appears to be stuck.

Some Criteria for Evaluating Counselor Primary Empathy

• Accuracy
• Aware of content
• Aware of feelings

- Well-timed
- Tentative
- Concise
- Primary, not advanced empathy

PROCESS

In the large group, students discuss what they learned from the role-play, what questions they still have about primary empathy and attending skills, and what they think about the role of these skills in genetic counseling.

Estimated time: 75 to 90 minutes.

Written Exercises

Exercise 1

Identify two or three situations in your own life where you have experienced disappointment, loss, grief, etc.

- Describe the situation, what you were feeling, thinking, doing.
- How did you cope with the situation?
- What resources (including other people) did you turn to for assistance?

INSTRUCTOR NOTE

- This exercise could be done as part of a journal, as a small reflection paper, or verbally with a dyad partner.
- Students should be encouraged to choose examples carefully so that they do not inadvertently disclose more than they intended.

Exercise 2

Using the following list of feeling words, generate four or five synonyms for each:

happy	anxious	embarrassed
sad	uncertain	withdrawn
angry	responsible	hopeless
scared	reluctant	rejected
confused	torn	uncomfortable

Example: Interested: curious, engaged, involved, invested.

Exercise 3

Using the following list of mild intensity feeling words, generate feelings that are at the moderate and extreme intensity levels for each one.

Mild Intensity	Moderate Intensity	Extreme Intensity
confused		
sorry		
nervous		
dissatisfied		
hesitant		
uncomfortable		
irritated		
glad		

Example: Mild = puzzled; Moderate = surprised; Extreme = amazed.

Exercise 4

Use the following client description to construct a dialogue between the genetic counselor and the client in which the genetic counselor uses only primary empathy responses. Formulate between 5 and 10 client statements and 5 and 10 counselor statements.

CLIENT DESCRIPTION

A professional couple, both about 25 years old, recently had their first baby. The baby was found to have phenylketonuria (PKU). The couple has just learned all about the diet and the risk of mental retardation.

Co Response:

Cl Response:

Co Response:

Cl Response:

Etc.

Exercise 5: Primary Empathy Exercise*

I. IDENTIFICATION OF FEELINGS

For each client statement below, list three to four possible feelings that the client could be experiencing when saying the statement. Choose feelings that are close to the surface of what the client is saying; do not move to advanced empathy (hidden feelings).

- If I'd known the baby would have hemophilia, I'd never have gotten pregnant in the first place.
- If my Huntington test is positive, then my life will be over.

- Why are you talking to me about an abortion? I've already told you that it's not an option!
- How can I even begin to tell my son that he has muscular dystrophy?
- I've already had three miscarriages; I'm not sure I can face another one.
- I don't know why I'm here. I just came because my doctor told me to.
- Can't you tell me anything for certain?
- You have no idea how hard it is to have another child who is mentally retarded!
- Since my mother died of breast cancer, my dad doesn't want me or any of my sisters to be tested.
- My sister refuses to give blood so I can figure out if I'm at risk.

II. MAKING CONTENT RESPONSES

Construct one concise content response for each of the client statements in part I. Write each response as if you were actually speaking to the client. Be sure that you have summarized the content.

III. MAKING AFFECTIVE (FEELING-ORIENTED) RESPONSES

Read each client statement again to yourself. Then write down one affective response for each. When you complete each response, read the client statement again. Does your response hit the most important feeling expressed? Does the response match the intensity of feeling expressed by the client?

Example: Client statement: I can't believe that I gave this disease to my child!

Genetic Counselor: I. Guilty, remorseful, responsible, ashamed

II. So, you feel like it's your fault?

III. It sounds like you feel ashamed.

[Hint: Read your content and affect responses aloud to be sure that they are concise, tentative, and convey the client's experience in your own words.]

Exercise 6

Engage in a 15- to 20-minute role-play of a genetic counseling session with a volunteer from outside the class. During the role-play, focus on primary empathy and attending skills. Audiotape the role-play. Next transcribe the role-play and critique your work. Give the tape, transcript, and self-critique to the instructor who will provide feedback. Use the following method for transcribing the session:

Counselor	Client	Self-Critique	Instructor
Key phrases of dialogue	Key phrases	Comment on your own response	Will provide feedback on your responses

---

*Adapted from Danish et al, 1980.

Annotated Bibliography

Baker, D. L. (1998). Interviewing techniques. In: Baker, D. L., Schuette, J. L., Uhlmann, W. R., eds. A guide to genetic counseling. New York: John Wiley & Sons, 55-74.
[Discusses the importance of identifying and addressing client affect.]
Bellet, P. S., and Maloney, M. J. (1991). The importance of empathy as an interviewing skill. Journal of the American Medical Association, 266, 1831-1832.
[Provides important arguments for the role of empathy in working with patients.]
Cormier, W. H., and Cormier, L. S. (1991). Interviewing strategies for helpers: fundamental skills and cognitive behavioral interventions (3rd ed.). Pacific Grove, CA: Brooks/Cole.
[Defines empathy and provides several specific examples and related exercises.]
Egan, G. (1994). The skilled helper (5th ed.). Monterey, CA: Brooks
[Distinguishes between primary and advanced empathy, provides examples and related activities.]
Kessler, S. (1999). Psychological aspects of genetic counseling: XIII. Empathy and decency. Journal of Genetic Counseling, 8, 333-344.
[Provides rationale for greater incorporation of empathy into the genetic counseling relationship/session. Offers specific examples of counselor responses that communicate empowerment, decency, thoughtfulness, positive reinforcement, and consolation.]
Pedersen, P. B., Ivey, A. (1993). Culture-centered counseling and interviewing skills: a practical guide. Westport, CN: Praeger Publishers.
[Puts empathy into a cultural context and provides illustrative examples.]
Punales-Morejon, D., Penchaszadeh, V. B. (1992). Psychosocial aspects of genetic counseling: cross-cultural issues. Birth Defects: Original Article Series, 28, 11-15.
[Describes several barriers to the use of genetic counseling services. Provides some useful information for the development of initial primary empathy for clients who differ from the genetic counselor.]
Rogers, C. R. (1992). The necessary and sufficient conditions of therapeutic personality change. Journal of Consulting and Clinical Psychology, 60, 827-832.
[Emphasizes the importance of empathy as a basic facilitative condition.]
Ridley, C. R., Lingle, D. W. (1996). Cultural empathy in multicultural counseling. In: Pedersen, P. B., Draguns, J. G., Lonner, W. J., Trimble, L. D., eds. Counseling across cultures (4th ed.). Thousand Oaks, CA: Sage Publications, 21-46.
[Defines cultural empathy, its origin and impact on helping relationships.]
Weil, J. (2000). Psychosocial genetic counseling. New York: Oxford University Press.
[Describes typical experiences of genetic counseling clients that can be helpful in gaining an empathic understanding; defines empathy and its role in genetic counseling.]

Chapter 5

Gathering Information: Asking Questions and Taking Client Genetic History

Learning Objectives

1. Distinguish between open-ended and close-ended questions.
2. Develop questioning skills through practice and feedback.
3. Identify a structure for client genetic history taking.
4. Develop skills obtaining information about client genetic histories through practice and feedback.

Obtaining Information from Clients

An essential component of genetic counseling is obtaining information about client situations in order to assess their reasons for seeking genetic counseling; the decisions, if any, that they wish to make; and the factors that are relevant to their situations. Questioning is an important skill for eliciting this type of information. This chapter first defines questioning skills and discusses effective and ineffective questioning strategies, and then discusses a specific type of information-gathering activity—collecting information about client histories.

Types of Questions

The most direct way to gather information from clients is by asking questions. Two major types of questions that are appropriate in genetic counseling are closed-ended and open-ended questions:

- Close-ended questions are questions that clients can easily answer with a "Yes," "No," or a one- or two-word response. Typically, closed questions begin with forms of the verb "to be": "Is it. . .," "Do you. . .," "Are they. . .?" (Danish et al, 1980). Closed questions either include an explicit or implied choice, such as "Are you going to have the test done?" (Hughes et al, 1997), or ask for specific information or details, such as "Do you have any children?"

(Hughes et al, 1997; Pedersen and Ivey, 1993). Thus, closed questions constrain the client's response.

- Open-ended questions are questions that clients cannot easily answer with a "Yes," "No," or a one- or two-word response. Typically open questions begin with such words as "How," "What," "Tell me about. . .," "I'm wondering about. . ." Open questions enrich the interview by inviting clients to freely express their views and experiences. They encourage clients to fill in the gaps with respect to their feelings, thoughts, and situations (Hughes et al, 1997; Pedersen and Ivey, 1993; Wubbolding, 1996). For example, you might ask, "How do you feel about the results of your test?" Open questions can encourage clients to disclose more fully, can elicit concrete, specific information, and can help you to better understand your client's situation (Geldard, 1989).

Consider the following examples of open and closed questions that might be asked during genetic counseling:

Closed question: Are you scared?

Open question: How do you feel?

Closed question: Are you concerned about what you will do if the test results are positive?

Open question: What do you think you might do if the test results are positive?

Closed question: Is your relationship with your husband a good one?

Open question: Tell me about your relationship with your husband (This response, although not grammatically a question, is still a question because it requests additional information).

In addition to being either open or closed, questions vary in their complexity. Sanders (1966) identified seven types of questions that differ in their degree of cognitive and affective complexity. His questions are based on Bloom's (1956) hierarchical taxonomy of educational objectives:

Memory questions require recall or recognition of information. Example: When did you have the miscarriage?

Translation questions require an idea to be expressed in different words. Example: Can you explain what you mean in another way?

Interpretation questions require generalizations of information. Example: What does a 1 in 10 chance mean for you?

Analysis questions require problem solving through critical reflection about available knowledge. Example: How can the ways that you've coped with loss in the past help you in this situation?

Synthesis questions require problem solving through original thinking. Example: Can you think of some ways to approach your family members about testing that you haven't tried yet?

Evaluation questions require value judgments. Example: Which of the options we've discussed fits the best for you?

Functions of Questions in Genetic Counseling

Wubbolding (1996) lists four functions of questions:

To enter the client's world by asking about client wishes, desires, perceptions, behaviors, etc. Example: How are you dealing with your fears about possibly having the breast cancer gene?

To gather information. Example: Have you had any previous pregnancies?

To give information. Sometimes questions subtly communicate information. Example: Tell me how you will deal with the insensitive comments from your coworkers. (This question suggests that the client has the strength to handle such comments.)

To help clients gain more effective control. Example: Please tell me whom you will contact if you begin to feel overwhelmed by the news I've just given you. (This question encourages the client to formulate an action plan.)

Questions set the stage for goal-setting (e.g., identifying what would be helpful for the client) and action (e.g., client decision making) by identifying and clarifying client problems, concerns, feelings, motives, and factors that affect their decision making (Pedersen and Ivey, 1993). The more information you have, the more valid your assessment of client goals, and the more helpful you will be in assisting clients in their decision-making processes.

Asking Questions Effectively

KNOW WHEN TO ASK A QUESTION

There are three times when a question can be particularly effective:

- When you have a clear reason for asking a question. The ultimate test of a question is whether it will be helpful to your client (Hill and O'Brien, 1999). Before asking a particular question, consider whether you would know what to say if your client asked why you wanted to know.

- When you want to gather more information and/or clarify client meaning. You may not have the same definitions as your clients for certain words and experiences.

- When you don't understand. One of the biggest mistakes you can make is to assume that you understand without checking out those assumptions with your client (Baker, 1998). Questions help to clarify your misperceptions. Another typical mistake, especially among beginning genetic counselors is acting as if you understand when you do not. Clients will know if you are only pretending to follow them (Martin, 2000).

Know How to Ask a Question

- Use questions strategically. Questions are very directive. For example, if your client says, "I'm upset that my family doesn't understand what I'm going through," you could respond, "Tell me more about your family"; "What do you mean by 'understand'?"; "What's it like to be upset?"; "What makes you think they don't understand?"; "What makes that so important to you?" Each of these questions would take the conversation in a very different direction. Be sure that you know which direction you want to go. Also, the indiscriminate use of questions will cause most clients to withdraw (Geldard, 1989). Inexperienced counselors tend to use more questions than experienced ones (McCarthy et al, 1977); with experience, you will appreciate the value of interspersing empathy and silence with questions. It is tempting to ask questions because they fill silences (something beginning counselors are uncomfortable with), and they demand that clients respond (Martin, 2000). You should be careful about relying too much on them.

- Be specific and comprehensive. When gathering information, ask about your client's thoughts, feelings, behaviors, and social systems (Cormier and Hackney, 1987). You should also request concrete examples (Hill and O'Brien, 1999).

- Be systematic. Stay with one topic before jumping to others. Follow up on content from your client's previous statement, or bring your client back to a topic if the client is topic-hopping. Begin with more general questions, and ones that are easier and less threatening to answer. Gradually move to more specific questions concerning more complex or more threatening issues.

- Keep questions simple. Ask one question at a time. Clients will be confused if you string several questions together in one response (Cormier and Cormier, 1991). For example, "What are you thinking and feeling about having an amnio, and what do you think you'll decide to do?" You should separate these into a series of questions.

- Avoid interrogating. Intersperse other types of responses with questions. For instance, follow up a question with an empathy response that clarifies or summarizes your client's answer.

Know What Type of Question to Ask

- Use both open- and closed-ended questions. Open-ended questions allow clients to express themselves more autonomously. For example, clients can choose to express things that are most important to them (Geldard, 1989). Open questions can be especially effective when first introducing a topic (Baker, 1998). Closed questions, as we stated earlier, allow you to gather precise information (e.g., details about family history), and they can keep the session from wandering off track (Pedersen and Ivey, 1993).

- Be sure to ask the type of question that you intended to ask. It is generally a mistake to ask closed questions when you want extended answers and to ask open questions when you want precise answers. Although in our personal lives, people who know us well will automatically provide elaborate responses to our closed questions, this is not the case in genetic counseling (Baker, 1998). Also, you need to clearly understand the difference between open and closed questions. Students often believe they have asked an open question "if they themselves assuredly intended to invite a narrative, even if their actual words were stated in a narrow or closed manner. However, clients often decline to amplify their replies if not specifically invited to do so with open questions" (Baker, 1998, p. 72).

- Use follow-up questions. If a client provides little or no response to your question, consider saying, "Tell me more about that" (Baker, 1998).

Follow These General Guidelines for Effective Questioning

- Avoid interrupting. Unless clients ramble excessively, allow them to finish sentences and thoughts. Ask clients to complete unfinished sentences.

- Allow clients to interrupt you. Usually, what they have to say is important and indicates that they are engaged in the process and willing to have a discussion with you.

- You can always backtrack. If the discussion shifts before you have gathered all of the information that you need, remember that you can always redirect your client. For instance, you could say, "Earlier, you were saying that. . . Can I ask you more about that?"

- Reinvite clients to share an experience. If your client provides only a minimal response to your open question, ask the question again. As Baker (1998) points out, "For many clients this will be the first time they have been invited to tell their story to a professional, and hesitance is understandable. Alternatively, they may be conveying that they don't feel ready to talk about the topic at this time. If the latter is your assessment, you may want to provide a summary of your thoughts, so that either party can return to this topic later" (p. 60).

- Use silence. Silences allow clients the time and space to consider your question and to formulate thoughtful responses. As we mention in Chapter 4, silence can be difficult to gauge. Try counting silently to 10 or 15 in order to allow sufficient time.

- Try other ways to obtain information. Remember that a good empathic reflection can lead clients to disclose a great deal of information (Martin, 2000). Their disclosure may not be as systematic as in response to a question, but clients do have more autonomy and control over what is discussed when you use a less leading response such as empathy. Monitor your questioning behavior to see if you are asking questions because you don't know what else to do, or because you are trying to avoid client feelings (Fine and Glasser, 1996). If this is the case, try to use other types of responses.

- Adapt your questioning to your client's needs and cultural background. Questions may come across as rude and intrusive to clients from some cultures (Pedersen and Ivey, 1993). Explain at the beginning of the session that you will need to obtain certain information and why you need to do so. Use words and sentence structures that are familiar to your clients (Pedersen and Ivey, 1993; Segall, 1986). Fisher (1996) offers these examples:

Some Southwest Native American tribes believe that "speaking about a deformity may give it the power to manifest itself in human form" (p. 78).

Some members of the Navajo tribe and others may not answer questions immediately (p. 83).

Some Southeast Asian clients will more readily answer questions such as the date of the last menstrual period when you are the same gender as they are (p. 118).

- Follow up with primary empathy. Summarize your client's response to your question (see Chapter 4 for a discussion of primary empathy).

- Remember that not all questions are questions. Responses that you state in a questioning tone are not necessarily questions. The primary goal of a question is to gather additional information. Empathy responses often are stated in a questioning tone, but their intent is to reflect the client's experience, not to gather new information.

Limiting Your Use of Open and Closed Questions

We cautioned earlier against the indiscriminate use of questions and stressed the importance of not interrogating clients. We elaborate on these points here because we believe that it is very important for beginning counselors to recognize both the potential benefits and limitations of questions. Two specific reasons to ask questions are to learn about clients and to cue clients about the types of information that you need. When you want your client to elaborate, an open question may be more effective (e.g., "What types of options do you believe you have open to you?"). If you need to efficiently gather specific information, then a closed question may be more effective (Hughes et al, 1997). Closed questions are particularly useful for gathering information for family and medical histories (e.g., "How many siblings do you have?").

Before asking questions, consider their potential impact. Open and closed questions may have different effects on clients and the counseling process. In mental health counseling, open questions tend to be perceived by clients as helpful, and they tend to encourage client disclosure (Elliott, 1985; Ehrlich et al, 1979; Hill et al, 1988), while closed questions tend to restrict client disclosure (Ehrlich et al, 1979; Hughes et al, 1997).

Avoid using a lot of questions prematurely. Because both open and closed questions are leading responses, when you use them, you direct the client towards certain discussion topics (Danish et al, 1980). When you ask questions you assume that you know better than your client what s/he needs to discuss. While this may be true to a certain extent, you cannot know everything a client needs to say, nor when and how the client needs to say it. You should be careful about using a lot of questions before you have established enough rapport to elicit open, honest client responses to your questions and before you have explored your client's concerns enough to be confident that your questions will move the conversation in relevant directions rather than going off on tangents or missing important areas.

Too many questions can lead to an interrogation in which you control the process, with the client becoming less rather than more communicative (Geldard, 1989). Bertakis and his colleagues (1991) found that patient satisfaction was highest when physicians communicated interest and friendliness and avoided behaviors that were dominating, such as excessive questioning. Overreliance on closed questions has been found to lead to lower client satisfaction during school counseling consultations (Hughes et al, 1997), during medical appointments with physicians (Bertakis et al, 1991), and during mental health counseling sessions (Ehrlich et al, 1979; Hill et al, 1988).

You must be sure that your questions are focused on your client's needs and reasons for seeking genetic counseling, and not out of your curiosity (Cormier and Cormier, 1991). For example, "So, what's it like to have a baby with a birth defect?" sounds like a question intended to satisfy your own curiosity. Compare this question with the following: "Please tell me how you are dealing with raising a child with cystic fibrosis." The latter question is more appropriately focused on the client's situation and needs.

Clients can become defensive if they are bombarded with a string of questions, especially if the questions appear to challenge something they have just said (Wubbolding, 1996). Consider the following example of excessive and challenging questions:

Client: I don't want to have a baby with Down syndrome.

Counselor: What do you mean?

Cl: I just don't think I could handle it.

Co: Are you telling me that you'd want to terminate the pregnancy?

Cl: Well, I'm not sure. . .

Co: Where does this feeling that you "couldn't handle it" come from?

Cl: I don't know what you mean.

Co: Well, is it coming from you or from your family?

Cl: Well, I guess from them.

Co: Do they have to live with the consequences of this decision, or do you?

Cl: Well, it will affect them, too.

Co: But can they really tell you what to do?

Cl:	No, I guess not.
Co:	So, let me ask you, what do you want to do?

In this example, the counselor overwhelmed the client with a series of questions that were quite presumptive and seemed to demand a certain answer. These questions raised a barrier between the counselor and client as the client became increasingly defensive and distressed.

Compare the previous example to this more appropriate interchange:

Cl:	I don't want to have a baby with Down syndrome.
Co:	Tell me more about that.
Cl:	I just don't think I could handle it.
Co:	So, you are afraid that you couldn't manage?
Cl:	Right. . ..I have to work full-time and I'd have no one to watch the baby. And I'm not sure I could give it all of the special care it would need.
Co:	What are your impressions of what a child with Down syndrome is like?
Cl:	[Client describes her perceptions.]
Co:	[Counselor affirms or corrects client perceptions, including a discussion of different levels of severity and then says:] What options have you considered if the test results indicate Down syndrome?

Wubbolding (1996) cautions against the use of questions that "mask" the counselor's opinions. For example:

Do you think you should be making this decision alone?

Do you think you should wait until you have the results of the amnio to confirm the ultrasound findings?

Do you think you should talk to your mother and find out more about her health history before having this test?

Although it may be appropriate at times to express your opinions, you should not disguise them as requests for more information.

Questions that You Generally Should or Should Not Ask

Although every client and genetic counseling situation is different, there is one critical question that you should consider asking most clients early in the session: "How can I help you?" or "Can you tell me what you're hoping to get out of genetic counseling?" A client's response to these questions can indicate her or his major goals for the genetic counseling session as well as give you some idea about how much your client understands about genetic counseling. However, there can be risks with these questions. For example, some clients can be overwhelmed if you ask these questions right away. They are not always sure of what they want and they may not even know why they are seeking genetic counseling (especially likely if they were referred). So, these questions will not be a magical way to get clients to open-up to you. You may need to modify this approach and first spend time describing what genetic counseling is and what they might be able to get from the session.

There is one type of question that you generally ought to avoid—"Why" questions. "Why" questions suggest that the person acted in a rational way when actually a great deal of human behavior is based on irrational, unplanned, unconscious forces, or is due to habit or ritual; your client will probably make up a reason, when there actually isn't a rational one (Krueger, 1994). Furthermore, "Why" questions imply judgment (e.g., "Why didn't you talk to your doctor about having the test?"), and the client may become defensive, feel guilty, or be offended (Egan, 1994; Geldard, 1989; Hill and O'Brien, 1999; Pedersen and Ivey, 1993). Any response to this type of question will be a rationalization or excuse (e.g., "Well, I didn't have the time to call and schedule an appointment with her") (Geldard, 1989). "Asking your client, 'Why do you feel that way?' can't be answered, doesn't go anywhere, and may well make your client defend himself. . ..If a why question is necessary, we find that the client feels better if you tell him why you are asking him why" (Fine and Glasser, 1996, p. 69). For example, "I'm asking you why you didn't have an amnio done because it might relate to your decision about this pregnancy." You can try to rephrase "Why" questions, for example, "What made you decide not to talk to your doctor about having the test?" Be aware, however, that rephrased questions may still sound judgmental.

Genetic History Taking: Constructing a Pedigree

We have stressed caution in your use of questions. Although questions can be a valuable tool for gathering information, they are only one of several skills that are important for effective genetic counseling. As we've stated previously, questions are most useful when you use them strategically, with a purpose that is clear to you and to your client.

Strategic questioning is particularly appropriate during the family history taking and pedigree construction phase of genetic counseling. This is a highly structured phase of the session in which you use questions to gather specific information from your clients. To manage client expectations, you should explicitly tell clients that you will ask a series of questions in order to understand their family history.

Definition of a Pedigree

The pedigree is the diagram that records the family history information, the tool for converting information provided by the client and/or obtained from the medical record into a standardized format. It demonstrates the biological relationships of the client to his or her family members through the use of symbols, vertical and horizontal lines, and abbreviations. When complete the pedigree stands as a quick and accurate visual record that assists in providing genetic counseling. [Schuette and Bennett, 1998, p. 27]

Papadopoulos et al (1997) describe the pedigree as "a clinical tool used for acquiring, storing, and processing information about family history, composition, and relationships" (p. 17). Pedigrees are also known as genograms, family trees, or genealogic charts. In this chapter, we use the term pedigree because it is the most commonly used term in genetic counseling and is more medically focused than relationship focused.

Family history taking is an essential component of genetic counseling because it provides "a basis for making a diagnosis, determining risk, and assessing the needs for patient education and psychosocial support" (Schuette and Bennett, 1998, p. 27). It is also a good way to begin to establish rapport with a client.

Functions of Pedigree Construction

The pedigree may serve several functions that impact both the process and the outcome of genetic counseling:

PROCESS FUNCTIONS OF A PEDIGREE

- It can foster counselor empathy and lead to greater rapport between the genetic counselor and client(s) (Bennett, 1999; Erlanger, 1990; McGoldrick and Gerson, 1985; Rogers and Durkin, 1984; Schuette and Bennett, 1998).

- It puts the client in the role of expert and the counselor in a one-down position; this can be especially important if clients feel out of control of their situations, or are mistrustful of health professionals (Erlanger, 1990). Clients may be more likely to view themselves as active participants in the etiology and management of their conditions (Stanion et al, 1997).

- Some research suggests that clients view pedigree construction positively, for instance they like to provide information, it makes them feel listened to, and the process eases their anxiety (Erlanger, 1990; Rogers and Durkin, 1984; Rose et al, 1999).

- It can help clients who are uncomfortable with open-ended questions to respond because questions are asked in a more systematic, matter-of-fact way (Papadopoulos et al, 1997).

- It provides a mechanism for considering information about patient risk and serves as a stimulus for discussing genetic risks, relevant tests, and further actions (Bennett, 1999; Rose et al, 1999; Schuette and Bennett, 1998).

- It provides an immediate illustration of the family's medical history that can be more easily updated, and important information can later be located more easily as compared to a narrative report (Papadopoulos et al, 1997; Rogers and Durkin, 1984; Stanion et al, 1997).

Outcome Functions of a Pedigree

- It allows for an assessment of the client's understanding.

- It offers an organized framework for seeing the big picture in the context of the whole family, which can free the client from feeling as if everything is her or his fault (Erlanger, 1990; McGoldrick and Gerson, 1985).

- It assists in client education, for example, providing information about whom in the family is at risk, about how the condition may vary in its expression (e.g., different ages of onset), and clarifies client misunderstandings (Bennett, 1999).

- It can offer the genetic counselor clues about the client's worldview, values, perceptions, lifestyle, religion, ethnic and cultural background, and family dynamics (Bennett, 1999; Erlanger, 1990; Schuette and Bennett, 1998).

- A great deal of information can be elicited in a fairly short amount of time (Erlanger, 1990; Rogers and Durkin, 1984). Some research indicates that formal history taking and pedigree construction can generate four times more family medical information than does an informal interview (Rogers and Durkin, 1984).

- It provides a valuable database for diagnosis and intervention (Bennett, 1999; Erlanger, 1990). The genetic counselor can gather information about the possible causes and effects of certain genetic conditions (Papadopoulos et al, 1997; Schuette and Bennett, 1998).

- A pedigree informs decisions about medical management (Bennett, 1999).

- It can illustrate inheritance patterns (Bennett, 1999).

- It helps in identifying reproductive options (Bennett, 1999).

- It distinguishes genetic risk factors from other types of risk factors (Bennett, 1999).

- In addition to medical information, a pedigree can provide information about social relationships, family crisis and loss, and family beliefs and misperceptions (Schuette and Bennett, 1998).

- It can potentially prevent the clinical onset of certain genetic conditions if it leads to early diagnosis and intervention, for example, vascular surgery for a child with Marfan syndrome (Papadopoulos et al, 1997), or early screening in familial adenomatous polyposis (FAP).

Psychosocial Aspects to Remember in Gathering Genetic Family History

- Introduce the process as taking a family history or determining the family tree rather than using a technical term such as pedigree, which can be intimidating (Stanion et al, 1997).

- Explain why you are constructing this family history, for example, to assist you in understanding and providing the most appropriate services you can for the client, and describe the process (a series of questions and drawing a diagram) (Stanion et al, 1997).

- Throughout the data collection, maintain a nonjudgmental, caring, open manner (Stanion et al, 1997).

- Begin with easy, factual questions to ease the client into the process (Stanion et al, 1997).

- Listen sensitively for clients' reluctance to admit to certain conditions because they feel guilty, fear stigmatization, or are embarrassed (Bennett, 1999). Pay attention to the emotional and cognitive impact that your information gathering is having on clients and slow your pace, if necessary (Stanion et al, 1997).

- Watch your terminology in order to be sensitive to the client. Some words may be frightening or offensive to the client, (e.g., "defect"; "abnormal"; "negative history" to refer to a history with no genetic condition—this is negative in our terminology but it is quite positive for the client (Bennett, 1999). Avoid medical terminology that the client would not understand.

- Respond empathically when the client reveals a traumatic experience. For example, when you take a pregnancy history, if a client reveals that one of her children died of Duchenne muscular dystrophy, you might say, "I'm sorry for your loss" (Bennett, 1999).

- Remember to be aware of cultural differences. Individuals from different backgrounds tend to have quite different explanations for the causes of genetic conditions and have different expectations about the impact of genetic conditions on themselves and on their familial and social networks (e.g., first cousins in a family from a Middle Eastern culture).

- Conduct the history taking and pedigree construction early in the session before moving to other aspects of counseling and evaluation (Schuette and Bennett, 1998).

- Make the history gathering process separate from the rest of the counseling. "If you ask informational questions at the beginning, it's important to let the client know by your behavior that the mood is going to change when this is completed. Put the protocol or file and pen away from you. Turn your chair slightly. Change your posture" (Fine and Glasser, 1986, p. 69).

Special Counseling Circumstances to Consider When Gathering a Genetic Family History

- Some clients may have preexisting beliefs about why there is a genetic condition in their family (e.g., fate, punishment for a sin, etc.) and they may not wish to be told otherwise. Cultural differences may lead clients to arrive at quite different conclusions about causality (Schuette and Bennett, 1998). For example, we have found that Mexican immigrants sometimes believe that health professionals give a genetic condition to their children or that birth defects are due to looking at a lunar eclipse while pregnant (McCarthy Veach et al, 2001).

- Some clients may become distressed when talking about the information you are eliciting (Papadopoulos et al, 1997). You should note distressing areas and return to these after constructing the pedigree (e.g., a pregnant patient who is upset over a previous miscarriage might have strong reactions to discussion of that pregnancy).

- Some clients may exhibit what Papadopoulos et al (1997) call the *Rashomon effect* (i.e., different family members see the same event differently). You must attempt to ascertain the most valid perspective when there is disagreement, or indicate on the pedigree that an issue is uncertain if the family cannot reach a common view.

- Clients may have misconceptions about inheritance that may affect what they hear during the session. It is important to be aware of them so that they can be addressed during the session. Correcting misinformation must be done with care and consideration of the impact. Clients may have lived many years believing that they are not at risk for the family disorder because of some misinformation passed on by family members. Some common misunderstandings about inheritance, such as the following, are discussed in more detail in Bennett's book, *The Practical Guide to the Genetic Family History:*

- Since there is only one affected person in the family, the condition cannot be genetic.

- If there is more than one affected person, it must be genetic.

- The mother did something during her pregnancy that caused the condition in the child.

- A patient will inherit a family condition because s/he looks like an affected family member.

- If all affected people in the family are of one gender, the other gender cannot be affected; the condition must be sex linked.

- Diseases skip generations, and birth order affects the risk.

Closing Comments

Questions can be particularly useful for gathering relevant family and medical histories. Since the family history is often one of the first activities to take place in a genetic counseling session, it provides an excellent opportunity to learn more about the client's concerns, perceptions, family relationships, and support systems. It is also a wonderful vehicle for establishing rapport. In the typical genetic counseling session, time is a major issue. This means that most questioning during this phase of the session is closed ended. However, particularly revealing client responses during this phase can be revisited later in the session and can greatly enhance counseling. Although family history gathering tends to become routine, it is a process that can set the tone and the framework for the session. History gathering requires a great deal of skill in order to ask the correct questions and to actively listen to what clients are telling you about themselves and their families.

Class Activities

Activity 1: Think-Pair-Share Dyads

Students individually consider the following questions and then discuss them with a partner:

- What is the role of questioning in genetic counseling?
- What are the potential benefits of asking questions?
- What are the potential costs of asking questions?
- Are there certain types of clients with whom you would be particularly cautious about asking questions?
- What are some reasons that clients might not answer your questions?

Estimated time: 10 minutes.

PROCESS

Dyads report on their discussion. The instructor summarizes major themes and presents any ideas that did not emerge from the dyads.

Estimated time: 15 minutes.

Activity 2: Brainstorming Questions

Present students with a brief genetic counseling client description/statement and ask them to generate all of the questions they can think to ask this client. This activity can be repeated several times with different client descriptions.

CLIENT STATEMENTS

- I'm afraid I'm going to get breast cancer.
- I was hoping that you could tell me my chances of having another miscarriage.
- My sister has a child with cystic fibrosis. I don't want that to happen to me.
- I want every test there is to make sure my baby is OK.
- My cousin has neurofibromatosis (NF) and I have some spots. My doctor thinks I have it, too.

Estimated time: 15 minutes.

Activity 3a: Low-Level Model

Instructor and a volunteer genetic counseling client engage in a role-play in which the counselor demonstrates poor questioning skills (e.g., closed questions, "Why" questions, strings several questions into one response, implies an opinion—"Don't you think that. . ."). Students should observe and take notes of examples of poor questioning.

Estimated time: 10 minutes.

PROCESS

Students discuss their examples of poor question asking. Then they discuss the impact of the counselor's poor skills on the client.

Estimated time: 10 to 15 minutes.

Activity 3b: High-Level Model

Instructor and a volunteer genetic counseling client engage in the same role play, but this time the counselor demonstrates good question asking skills, as well as good empathy and attending. Students should observe and take notes of examples of good questioning, empathy, and attending.

Estimated time: 10 minutes.

PROCESS

Have students discuss their examples of good counseling skills, especially focusing on question asking. Then discuss with students the impact of the counselor's good skills on the client.

Estimated time: 10 to 15 minutes.

Activity 4: Triad Exercise

Three students practice using questions, primary empathy, and attending skills in 10- to 15-minute role-plays taking turns as counselor, client, and observer. Allow 10 minutes of feedback for each role-play.

CRITERIA FOR EVALUATING COUNSELOR QUESTIONS

Concrete and specific (asks for examples)

Systematic (questions seem planned)

Comprehensive (covers thoughts, feelings, behaviors)

Uses silence

Avoids interrupting

Avoids use of "Why" questions

Follows up questions with primary empathy

Uses open questions where possible

PROCESS

In the large group, discuss what students learned from the role plays, what concerns or confusion they still have about questioning skills, and what they think about the utility of questioning skills in genetic counseling.

Estimated time: 75 to 90 minutes.

Activity 5: Brainstorming Family History Content Areas

Students generate areas in which genetic counselors would question a client in order to obtain a family history and construct a pedigree. Instructor records their ideas on the board and fills in any areas that they miss.

Estimated time: 15 to 20 minutes.

Activity 6: Constructing a Pedigree

Instructor interviews a volunteer from class (or from outside of class) to gather a family history and draws a pedigree on the board or on an overhead as the volunteer provides information. Instructor explains why s/he is using certain symbols, notations, lines, etc. as the interview progresses.

Estimated time: 45 to 60 minutes.

Activity 7: Model Family History Taking

Instructor and a volunteer from outside of the class engage in a simulated genetic counseling session in which the instructor gathers family history. Students observe and take notes on family history. They also record examples of the genetic counselor using good question asking skills, empathy, and attending. This session should be audiotaped for students to use in written Exercise 5, below.

Estimated time: 30 minutes.

PROCESS 1

In small groups, students summarize the family history information according to major categories (identified in Activity 5, above), or using categories provided by the instructor. Possible categories include:

- Biological relationships
- Significant life events (births, deaths, marriages, divorce, etc.)
- Role of extended family members (close and supportive, distant and uninvolved)
- Experience with health care delivery (close family member with chronic illness vs. no health issues in close family members)
- Experience with diseases
- Educational level

Estimated time: 20 minutes.

In a large group, instructor has students provide their group summaries for each category (a different small group can provide information for each area, and the other small groups can add or modify).

Estimated time: 20 minutes.

PROCESS 3

In a large group, discuss student observations about the genetic counselor's use of good questions, empathy, and attending.

Estimated time: 15 minutes.

Activity 8: Dyads: Pedigree Construction

Pairs of students practice constructing pedigrees for each other.

PROCESS

In a large group, discuss any confusion students have, clarify pedigree construction process, etc. Talk about what they perceive to be the benefits of constructing a pedigree.

Estimated time: 75 minutes.

INSTRUCTOR NOTE

• This activity can be extended so that each student takes pedigrees from several other students. Given time constraints, it could be done as an "outside" exercise rather than an "in class" activity.

Activity 9: Dyads: Constructing and Interpreting Pedigrees

Students construct their own pedigrees.* Dyads exchange pedigrees in class. Next, each student writes a narrative of the family history based on the pedigree. The dyad members discuss the narratives and pedigrees until both are clear and accurate.

Estimated time: 60 minutes.

INSTRUCTOR NOTE

* This part of the activity could be done outside of class.

Activity 10: Dyads: Correcting Misinformation

Pairs of students discuss ways of correcting some common client misperceptions:

1. Since there is only one affected person in the family, the condition cannot be genetic.

2. If there is more than one affected person, it must be genetic.

3. The mother did something during her pregnancy that caused the condition in the child.

4. A patient will inherit a family condition because s/he looks like an affected family member.

5. If all affected people in the family are of one gender, the other gender cannot be affected; the condition must be sex linked.

6. Diseases skip generations.

7. Birth order affects the risk.

PROCESS

In a large group, members can discuss and give feedback about their responses.

Estimated time: 60 minutes.

INSTRUCTOR NOTE

• This activity could be done in the clinical setting with the clinical supervisor providing feedback about responses or in a written exercise.

Written Exercises

Exercise 1*

Rewrite each of the following counselor closed-ended questions, turning them into open-ended questions.

• Do you understand this information?
• Do you have any questions?
• Are you upset?
• Are you OK with having a child with Down syndrome?
• Does this test make sense to you?
• Do both of you agree about having this test?
• Does your fiancé know about this disease in your family?
• Do you want any more children?

Exercise 2

Refer to the counselor/client dialogue where the counselor's excessive use of questions led to the client feeling defensive. Create a similar dialogue between a genetic counselor and client. First write the interchange with the counselor

---

*Adapted from Geldard, 1989.

asking excessive and challenging/judgmental questions. Then rewrite the interchange, using a combination of counselor questions, primary empathy responses, and silence, as appropriate. (Of course, client responses would be expected to change in response to these more effective counselor interventions.)

Exercise 3*

Write one question for each of Sanders's (1966) seven types of questions. Write your questions as if you are actually asking them during a genetic counseling session.

Memory question:

Translation question:

Application question:

Synthesis question:

Analysis question:

Evaluation question:

Exercise 4

Students audiotape an interview in which they gather a family history from a volunteer. Next they construct a pedigree based on the information obtained in the interview. Students submit their tapes and pedigrees for evaluation.

Exercise 5

Students construct a pedigree using the audiotape and notes from the simulated history taking session (see Activity 7) conducted by the instructor.

Annotated Bibliography

Baker, D. (1998). Interviewing techniques. In: Baker, D. L., Schuette, J. L., Uhlmann, W. R., eds. A guide to genetic counseling. New York: Wiley-Liss, 55-74.
[Provides a detailed discussion of basic helping skills relevant to genetic counseling. Describes the different types of genetic counselor questions and their functions in genetic counseling.]
Bennett, R. L. (1999). The practical guide to the genetic family history. New York: Wiley-Liss.
[A comprehensive examination of the use of pedigrees for genetic issues. Includes specific content areas for different types of genetic conditions and describes methods and techniques for family history taking and pedigree construction.]
Dillon, J. T. (1990). The practice of questioning. New York: Routledge.
[Discusses the purposes of using questions in counseling and therapy interviews.]

---

*Adapted from Pedersen and Ivey, 1993.

Fisher, N. L., ed. (1996). Cultural and ethnic diversity, a guide for genetics professionals. Baltimore: Johns Hopkins University Press.
[Discusses possible negative reactions of Southeast Asians to genetic information that is couched in terms of "bad news." This book provides additional information about the impact of diversity on genetic service provision.]
McGoldrick, M., Gerson, R. (1985). Genograms in family assessment. New York: W. W. Norton.
[A classic book on the nature and use of genograms. These authors are credited with introducing genograms into counseling settings.]
Putnam, S. M., Stiles, W. B. (1993). Verbal exchanges in medical interviews: implications and innovations. Social Science and Medicine, 36, 1597-1604.
[The authors analyzed medical interviews and determined that physician question asking occurs throughout the interview, but it is predominant in the history-taking session. Discusses the role of closed question in diagnosis and clinical hypothesis formation.]
Rogers, J., Durkin, M. (1984). The semi-structured genogram interview: I. Protocol, II. Evaluation. Family Systems Medicine, 2, 176-187.
[Contains examples of questions that could be asked to elicit information about family health and psychosocial functioning.]
Sanders, N. M. (1966). Classroom questions: what kind? New York: Harper and Row.
[Describes seven types of questions that range in complexity. These include questions that require recall or recognition and questions that require evaluations or judgments.]
Schuette, J. L., Bennett, R. L. (1998). Lessons in history: obtaining the family history and constructing a pedigree. In: Baker, D. L., Schuette, J. L., Uhlmann, W. R., eds. A guide to genetic counseling. New York: Wiley-Liss, 27-54.
[A thorough description of family history taking and pedigree construction in genetic counseling. Contains several useful diagrams of the types of symbols and notations that genetic counselors would use with different types of genetic clients and genetic conditions. Offers practical tips on history taking and pedigree construction.]
Segall, M. H. (1986). Culture and behavior: psychology in global perspective. Annual Review of Psychology, 37, 523-564.
[Discusses how to modify questions so that they are more culturally appropriate.]

CHAPTER 6

Structuring the Genetic Counseling Session: Initiating, Contracting, Ending, and Referral

Learning Objectives

1. Describe activities for initiating the genetic counseling session.
2. Define contracting and describe steps in the goal-setting process.
3. Describe genetic counselor activities for ending the session and the relationship.
4. Identify referral strategies for effective follow-up.
5. Develop skills at initiating, contracting, ending, and referral through practice and feedback.

This chapter discusses four components of the genetic counseling session: initiating the session, contracting (setting goals), ending the session/relationship, and making referrals.

Initiating the Genetic Counseling Session

Close your eyes and imagine that you are about to see your first genetic counseling client. What are you feeling? What are you doing to prepare for your first encounter? Do you have a clear picture of how you will begin? What is the first thing that you will say or do? Now ask yourself how your clients may feel about coming to genetic counseling and what they might say or do.

Many people have never heard of genetic counseling prior to becoming genetic counseling clients. They may be anxious, confused, frightened, and disoriented about the relationship they are about to enter into with you. It is important that you try to alleviate their discomfort about genetic counseling by helping them understand what will happen in this relationship. There are several steps that you can take to set the stage for genetic counseling. We have organized these steps according to three phases identified by Cormier and Hackney (1987) for mental health counseling: preparation, introduction, and agenda setting.

Preparation

REVIEW CLIENT RECORDS

Review any available client records prior to the genetic counseling session. If information from a referring physician is missing, you should attempt to acquire it prior to seeing the client: "There is nothing more distressing to a client than the professional who is unfamiliar with recorded information that the client knows exists" (Fine and Glasser, 1996, p. 29). Reviewing the records not only better prepares you to assess client goals for genetic counseling, but also indicates to the client that you are interested and respectful enough to take the time to do this preparation.

ARRANGE THE COUNSELING ENVIRONMENT

Surroundings are an important aspect of setting the overall tone. Keep your office neat, uncluttered, and inviting.

- If possible, have chairs of approximately equal size and comfort available and facing each other (Martin, 2000).
- Position your chair so that you face your clients and move any desks so that they are not between you (Fine and Glasser, 1996). Have a box of tissues available.
- Turn off or turn down the telephone and post a Do Not Disturb sign on the door.
- If you carry a pager, turn it to vibrate mode and warn your client if you are expecting a page during the session.
- If you intend to audiotape the session, place the recorder in an unobtrusive spot, but do not hide it. Set up the equipment and try it out to be sure that it records properly.

PREPARE YOURSELF TO BEGIN

- Minimize distraction (Fine and Glasser, 1996). Be sure that your manner of presentation is not distracting. Wear clothes that are appropriate for your setting and for the clients with whom you are working. In genetic counseling, it is not appropriate to wear casual clothing such as jeans, hiking boots, and shorts. You should avoid clothing that could be regarded as provocative—short skirts, sheer blouses, low-cut shirts, thong-type sandals. Think about the type of impression that you wish to convey, and then dress and behave accordingly. Also, as we discussed in Chapter 3, work on reducing or eliminating personal habits that could be distracting (e.g., twisting your hair, playing with jewelry, excessive use of filler words such as OK, right, uh-huh).
- If you have time between sessions, take a moment to psychologically prepare for the next session. Sit in a quiet room, take some deep breaths to calm

yourself, try to put aside extraneous thoughts, and focus on the client(s) whom you will meet in a few minutes. Visualize how you will greet your client, and what you might say at the beginning and end of the genetic counseling session.

- Begin on time, if at all possible, so as not to keep anxious clients waiting.

INTRODUCTION

- Greet your clients and escort them to the room. Remember that there are cultural variations regarding touch. For example, some individuals from the Middle East do not shake the hands of members of the opposite sex. We recommend that you shake hands only if the client initiates this behavior.

- Introduce yourself by first and last name. There is no clear protocol about how to address clients, but your institution may have one. Also, if clients are older than you, it may be appropriate to address them more formally (Mr., Ms., etc.). Once in the room, ask clients what they would like to be called and how to pronounce their names (Cormier and Hackney, 1987). There are cultural differences in how individuals wish to be addressed. For example, some older immigrants from Asia or conservatively stratified societies may wish to be treated more formally rather than being called by their first names (Ishiyama, 1995; Sue and Sue, 1990).

- Do not insist that clients address you in a certain way. For example, some clients may not be comfortable using your first name. Pay attention to any changes in the way a client addresses you, as they may indicate a change in your relationship, either toward or away from trust and comfort (Fine and Glasser, 1996).

- Allow your clients to choose the chairs in which they sit, and wait for them to be seated first so that you don't communicate that you are rushing or directing them (Fine and Glasser, 1996). When your clients are couples or families, where they position themselves provides important clues, for instance, about power dynamics and degree of closeness among family members. So pay attention to who directs the action, who takes the most prominent seat, who sits closest to you, who sits next to whom, who speaks first and who speaks after whom, whether individuals introduce themselves or are introduced by others, and how they refer to each other (Fine and Glasser, 1996). Additionally, you can gain important clues about individual clients by how they position their chair. For example, do they move it closer to or further away from you? Do they pick a chair that is close to or farther away from yours?

- In general, you should limit your small talk to an amount that helps your client relax a bit. A small amount of social conversation can be beneficial for individuals from certain ethnic backgrounds. For instance, some Asians, Native Americans, Hispanics, and African Americans may prefer a brief period of social conversation before proceeding to more intimate topics (Fine and

Glasser, 1996). In our experience, this is true of most clients regardless of their ethnic background. Too much small talk, however, is usually due to your own discomfort rather than for the client's benefit.

Agenda Setting

- Explain the genetic counseling process to your client (see Chapter 11 for a discussion about informed consent for the genetic counseling session). As part of this process, you could ask clients to describe their understanding of what will happen in the session. For example, "What was your doctor able to tell you about your appointment today?" If clients are too confused about the process to answer this question, then before moving to the medical/family history, you might say, "Is there anything in particular that you are concerned about or would like to ask me about?"

- Next, move to an opening question, such as, "What would you like to focus on?" (Fine and Glasser, 1996; Hill and O'Brien, 1999). One caution, however, is that questions can have different meanings cross-culturally. For example, some Native Americans regard questions as rude. We recommend that when explaining the genetic counseling process you state that you will ask questions and tell clients why you need to do this.

- Try a slower pace with clients from cultures where discussion of private issues and expression of negative emotions is discouraged (Ishiyama, 1995; Sue and Sue, 1990).

- If a client is visibly distressed, then you might begin by saying, "You seem very upset" or "It's hard for you to begin." If the client is a reluctant participant, for example, referred against her or his wishes, begin with an acknowledgment of this situation (Fine and Glasser, 1996): "I realize that you may be here because you were sent by _____, and you may be feeling uncertain [or uncomfortable, or angry] right now." Then wait for your client to respond. Do not try to defend or justify genetic counseling. Instead you might say, "I understand that it is not completely your choice to be here. But since you are *here, could we talk about whether there is anything we might do today that would be helpful for you?*"

- If you elect to take notes during the session, explain to your clients why you are doing so. Hold the paper so that they can see it if they wish. Be aware that you may lose valuable nonverbal cues because you are not able to consistently look at your clients. Taking notes unobtrusively, while simultaneously paying close attention to your client, is a skill that requires a fair amount of practice. You may wish to practice this type of note-taking outside of the genetic counseling session (e.g., role-playing with a friend). You will need to become proficient enough that your clients feel as if they have your full attention throughout the genetic counseling session (Fine and Glasser, 1996).

- Consider developing and using an interview checklist outlining the areas that you wish to cover during a session. Vriend and Kottler (1980) advocate using checklists in mental health counseling and argue that they do not interfere with the counseling session and that clients regard them as evidence of the counselor's thoroughness and attentiveness. A checklist may also serve as a stimulus for supervision (i.e., discussion of which topics were and were not covered, and which topics were most problematic for you). A personal checklist that best meets your style and criteria for what should take place during a session may help you proceed in an unhurried, but efficient manner and may also be helpful for constructing postsession notes (Vriend and Kottler, 1980). When using a checklist, remember that you will likely vary the order in which you raise topics, and you may not cover all topics with all clients. Also, as with note-taking, you should explain why you are using a checklist and allow the client to see it. (See Appendix 6.1 for a list of topics that you might include on a genetic counseling checklist.) Another option is to use a *mental* checklist.

- If you plan to tape the session, ask your client's permission. Present the purpose of taping in a matter-of-fact way (e.g., "I'm doing this in order to receive supervision on my counseling skills") and assure your clients of the confidentiality of the tape (Fine and Glasser, 1996; Martin, 2000). We recommend that you tell your clients that you will keep the tapes until the end of your clinical rotation and erase them at that time. If a client is highly resistant to taping, offer to turn off the recorder if s/he wishes you to do so at any point during the session, and then do it! Finally, secure your client's written permission to tape (Fine and Glasser, 1996).

- At any time during the session, if someone comes to the door, step outside and close the door to protect client privacy. Handle the interruption quickly, come back to your client, apologize, and briefly summarize what you were talking about before being interrupted (Fine and Glasser, 1996).

OBSTACLES TO INITIATING THE SESSION

- Clients typically enter genetic counseling without knowing exactly what to expect or what will be expected of them. They may feel uncertain, vulnerable, or even embarrassed. By describing what will happen and by conveying a caring attitude, you can help them adapt to the situation (Cormier and Hackney, 1987).

- Some task-oriented clients launch right in as though they know exactly what genetic counseling is all about - showing little evidence of uncertainty or anxiety. However, you should not assume that you've established rapport that quickly. Issues of trust, safety, and respect likely will arise later as clients begin to watch for your reactions. One way to address such uncertainties is to provide informed consent by describing the genetic counseling experience at the very beginning of the session (cf. Jacobson et al, 2001).

- Not all clients will be forthcoming about their situations: "Despairing, frightened, or angry clients are often less willing or able to expose certain aspects of their problems, because doing so risks injury to their precariously maintained self-esteem" (Marziali, 1988, p. 24). In Chapters 3 and 4 we discussed ways in which you can use attending and empathy skills to build rapport and trust that facilitates greater client self-disclosure. In addition,

Marziali (1988) identifies several counselor factors that strengthen the counselor-client alliance and lead to more positive outcomes in mental health counseling: commitment to helping the client, encouraging self-reflection in the client, and liking the client. Because these factors also seem relevant for genetic counseling relationships, we recommend that you monitor your reactions and behaviors to determine the extent to which you are committed to and like your clients, and how much you are encouraging client reflection.

Contracting

Definition of Goals and Their Function

Imagine yourself starting out on a vacation with a friend. The two of you are driving down the road, engaged in a lively conversation, as you head for Canada. However, about 30 minutes or so into the conversation, your friend looks around at the road signs and says, "Wait! I though we were going to Florida!" You slam on the brake. What happened in this situation? Evidently the two of you failed to discuss your intended destination; you did not develop an itinerary or road map for your journey.

A similar situation is likely to arise if you and your clients do not arrive at explicit and compatible goals for your session/relationship. Without a road map, sooner or later you'll have to pull over to the side of the road. Goals are the road maps that bring focus and direction to the session and help to structure the relationship: "Goals help both the counselor and client to specify exactly what can and cannot be accomplished through counseling" (Hackney and Cormier, 1996, p. 104). Goal-setting encourages clients to be clear about what they want to accomplish in a genetic counseling session. Furthermore, goals can help clients feel motivated to take action, and they allow both you and your clients to evaluate the effectiveness of the genetic counseling session and relationship.

The importance of explicitly stated and agreed-upon goals is illustrated in a study of concordance between genetic counselors' and clients' views of the nature/type of client concerns and the level/severity of their concerns. Michie et al (1998) found that genetic counselors were sometimes inaccurate in judging client concerns: "When there was not concordance, counselors were more likely than patients to think the patients' main concern was to get information or to find out about their risk status" (p. 228). Concordant sessions tended to emphasize more emotional issues and resulted in greater client satisfaction with the information received, and greater satisfaction with the extent to which their expectations were met.

Goals represent an individual's aims or purposes. In genetic counseling, they represent the client's and the counselor's expectations. Hackney and Cormier (1996) distinguish between process goals and outcome goals in mental health counseling. In general, "Process goals affect the therapeutic relationship. Outcome goals affect the results of counseling" (Hackney and Cormier, 1996, p. 121). Process goals refer to the conditions necessary to establish the relationship (e.g., being nondirective, demonstrating good attending behaviors, etc.). Within genetic counseling, process goals tend to be fairly general, applicable to all genetic counseling relationships. They are primarily your responsibility to accomplish, and they are not usually verbalized to clients. In contrast, outcome goals are unique to each client and each situation. They are more specific, and you and your clients share a mutual responsibility for their establishment. Outcome goals may change as the genetic counseling relationship progresses; therefore, a certain amount of flexibility is necessary in setting and sticking to goals.

What makes a goal feasible? Cavanagh (1990) identifies several criteria: it is specific, realistic, and mutually agreed upon; it defines the conditions necessary for reaching a desired outcome (e.g., making a decision, gaining genetic information, etc.); it is compatible with client and counselor values; and it is qualified, that is, it tends not to have an all-or-nothing quality. For instance, a client might say, "I want to know for certain that my child is OK." This is a very difficult, if not impossible, goal. Usually this type of certainty in genetic counseling cannot be achieved because of the complexity and the limits of genetic knowledge and testing (McCarthy Veach et al, 2001). A feasible goal is also open to revision as you and your clients reach new understandings (Martin, 2000).

The goals that are established cannot be more specific than either the counselor's or the client's understanding of the problem (Hackney and Cormier, 1996). So effective goal-setting requires solid attending and empathy skills, and good information-gathering skills. It also requires good inferential skills: "You must be able to listen to often vague client statements about what they hope to accomplish and 'read between the lines' of those messages" (Hackney and Cormier, 1996, p. 33). Most clients will be general and will tend to talk about goals in problem language. Your challenge is to reframe these statements into specific, positive goal statements. For example, a client says, "I don't want to make the wrong decision." You can reframe this goal as, "You want to learn about genetic risk factors, weigh your available options, and reach a decision based on that knowledge." Or, a client says, "I don't want to feel scared about my genetic status anymore." You might reframe this goal as, "You want to pursue genetic testing in order to see whether or not you carry the gene for muscular dystrophy." Sometimes you will need to use advanced empathy and mild confrontation to help clients set realistic goals. For example, it is not a realistic goal for a tearful prenatal patient who wanted the pregnancy to say, "I want to feel happy about my decision to terminate my pregnancy." You might say, "I wonder if it would ever be possible for you to feel happy about this. Perhaps you are saying that you want to feel confident that you made the best decision possible with the information we have?"

Strategies for Setting Goals

- Get a statement of your client's concerns or reasons for coming to genetic counseling.

- Notice client nonverbals and try to use that information to understand the client's emotional state.
- Translate problems into specific statements of goals toward which the genetic counseling will be oriented.
- Try for client agreement to work on these goals.
- If appropriate, arouse some hope in the client about what genetic counseling can provide.
- Ask the client, "What is your understanding of what we will be doing?" Then clarify any misconceptions.
- Try to establish both immediate and longer range goals, if appropriate (Cormier and Hackney, 1987). For example, an immediate goal would be, "To learn the occurrence risk rates for a genetic condition." A longer range goal would be, "To make a decision about whether to be tested."
- Try to establish goals that build on the client's resources and assets (Cormier and Hackney, 1987). For example, a client with strong religious beliefs might want to set a goal of consulting with a religious leader for assistance with the decision-making process.

Obstacles to Goal Setting

Danish and D'Augelli (1983) identify four major obstacles or roadblocks to client goal-setting and attainment, which we define in Table 6.1, along with examples and possible genetic counselor interventions.

These obstacles, either alone or in combination, can prevent clients from

TABLE 6.1. Roadblocks to Client Goal Attainment.

Roadblock	Client Example	Counselor Intervention
Lack of knowledge	Client does not know that there is a risk of passing a gene onto her child	Provide client with genetic risk information
Lack of skills	Client does not know how to approach his family members to persuade them to participate in genetic testing	Practice various scenarios for approaching family members
Fear of risk-taking	Client fears that she could not handle a positive test result	Discuss fears and refer for mental health counseling if appropriate
Lack of social support	Client has no supportive family members or friends, no religious/spiritual base, etc.	Refer to a support group or to mental health counseling

achieving their desired outcomes. The genetic counselor needs to assess client roadblocks and take steps to reduce or remove them.

Here are other examples of specific obstacles to setting and achieving goals:

- Some clients and genetic counselors may lack experience stating problems in positive, goal-oriented terms.

- The client really does not want to reach a particular outcome (e.g., by avoiding setting a goal of deciding whether or not to have prenatal testing, the client may believe she can make the issue go away).

- The client may be trying to set goals that actually belong to someone else (e.g., "My doctor wants me to have this test"; "My parents want me to terminate the pregnancy"; "Can you imagine what my neighbors would say if they knew I wanted to have a baby that might be retarded!"). Of course, if a client is attending genetic counseling primarily to satisfy another health care professional, or is going through the motions in order to be able to have testing, then the goals will be quite limited and not as mutual as you might like. In such situations you might say, "I know you would rather not be here. However, since you are, I'm wondering if there's anything that might be beneficial to you. Is there anything you might want to discuss?"

- Clients may resist goals that they perceive as being forced onto them, either by you or by someone else. For instance, McCarthy Veach et al (1999) found that some former prenatal clients were dissatisfied with their genetic counseling because the genetic counselor insisted on presenting termination as an option after they had explicitly stated that it was not an option for them. Clearly, these genetic counselors and clients were at odds over the goal of discussing all available options. Remember, you don't always have to go into detail about every option merely because you think that you need to cover all of the bases. It is important to respect your client's views and feelings.

- Cultural worldviews that outcomes are due to chance, fate, God's will, etc., may not be compatible with self-directed goal-setting. Clients with such worldviews may have difficulty seeing the value in setting goals. Nevertheless, you might say, "You have made the decision to come for genetic counseling. So I assume you believe there is something we could do that might be useful for you. How would you like to spend this time together?"

- Sue and Zane (1987) recommend three strategies for establishing goals when counselor and client cultural backgrounds differ:

  - Conceptualize client concerns in ways that are consistent with the client's culture.

  - Use interventions and identify strategies for goal attainment that fit with the client's culture.

  - Identify counseling goals that are concordant with the client's goals.

As part of this process, counselors may need to explicitly acknowledge that there are ethnic and cultural differences between their clients and themselves (Carlos Poston et al, 1991).

- Some clients come from cultures that do not have a future time orientation, and therefore goals should be linked less to dates and more to social or natural events (Brown, 1997). Also, Western views of change usually are linked to acting upon one's environment and taking control of one's situation, whereas for some clients from other cultures, change is regarded as establishing harmony within the family or tribe and learning to appreciate the ways things are and one's place in this reality (Brown, 1997). Clients who hold Eurocentric views will tend to take a goal-oriented, self-expressive approach to dealing with their problems as will many African-American and Asian-American clients, while Hispanic-American clients may tend to take a wait-and-see approach, and American Indians may prefer controlled self-expression characterized by thoughtful, rational, carefully controlled responses: "One implication of this value is that different groups may take longer to consider the problem and will have different propensities for action" (Brown, 1997, p. 34).

- Some clients lack an ability to conceptualize the void between where they are now, today, and where they would like to be (Hackney and Cormier, 1996). For example, a client who says, "I would never be able to make up my mind about terminating my pregnancy" fails to recognize that she might be able to arrive at a decision through a series of smaller steps (e.g., gather information; have the test; consult with genetic counselor and family members, etc.). These clients are aware of where they are currently and where they'd ideally like to be, but they are unable to visualize what they would need to do to get from here to there. You could ask, "What are some things you need to do in order to make this happen? What will be your first step?" (Cormier and Hackney, 1987).

- Some clients may lack a clear awareness of their values, desires, priorities, etc. (Hackney and Cormier, 1996), or they may be in conflict (e.g., wishing to determine if their child has Fragile X syndrome, but feeling responsible for this condition). You might address client confusion and ambivalence by acknowledging it (Cormier and Hackney, 1987). For example, "You don't have to come up with a plan right now. Would you like to take a few minutes to think about it?" You might try advanced empathy to identify the conflict. For instance, "I wonder if your indecision about Fragile X syndrome testing is due to your feeling responsible for your son's condition?"

- When your clients are couples or families, you must simultaneously take into account several individuals' wishes, desires, and needs (Martin, 2000). This can make goal-setting difficult as their interests may be in conflict. Indeed, one of the major challenges experienced by genetic counselors, physicians, and nurses when their patients have genetic concerns involves disagreements among family members about what to do (McCarthy Veach et al, 2001). In such situations you might say, "My goal is to assist you in finding the most satisfying solutions for all of you" (Martin, 2000). Then you should ask each individual to express her or his wishes. One possible exception would be situations in which it is a cultural practice for one family member to speak for another.

Goals of Genetic Counseling

Walker (1998) discusses the definition of genetic counseling provided by the American Society of Human Genetics (ASHG) in 1975. Within this definition, six possible types of outcome goals are evident:

- Comprehend medical facts including diagnosis, probable course of the condition, and available management.
- Understand the way heredity contributes to the condition and recurrence risks of specified relatives.
- Understand alternatives for dealing with recurrence risk.
- Choose a course of action that seems appropriate to the client in light of her or his risk, family goals, and personal ethical and religious standards.
- Act on the chosen course of action.
- Make the best possible adjustment to the disorder or risk of the disorder in an affected family member.

These goal statements, particularly themes involving choice of action and adjustment to the disorder or to the risk, illustrate how unique and individualized genetic counseling outcomes will be for your clients.

Genetic Counseling Endings

In genetic counseling, there are two types of endings: session endings and relationship endings. In cases where you have only one contact with a client (e.g., many prenatal genetic counseling cases), these endings occur simultaneously. In other cases, you may have contacts that extend over a period of years (e.g., working in a specialty clinic such as a muscular dystrophy clinic, where clients are seen for ongoing care; clients who return with subsequent pregnancies; BRCA gene families, etc.). In this section we offer suggestions for effectively managing both types of endings.

Guidelines for Effective Endings

In general, you set the stage for successful endings from the outset by describing the nature of the contacts you anticipate having with your clients (e.g., number of sessions, session length, follow-up contacts, etc.). To set the stage, we recommend the following:

- Educate clients. Explain how much time you have together at the beginning of genetic counseling so that it doesn't come as a surprise at the end (Kramer, 1990).

- Keep track of time. A clock on the wall behind your clients will provide you with an unobtrusive way to monitor the time. Also, keep a clock visible for your clients so that they can track the time as well.

- Prepare clients for the end of a session. About 10 minutes before the end of the session steer the conversation toward its conclusion (e.g., "There are only 10 minutes left, and I wanted to be sure that we had enough time to talk about..." or "There are only 10 minutes left, and I was wondering if there is anything else that you wish to discuss"; or "We have about 10 more minutes. Is there anything else that you would like me to know?"). This 10-minute warning allows clients to compose themselves if they are upset, to reflect on what has been discussed, and to wrap up any loose ends or unfinished issues (Geldard, 1989; Hill and O'Brien, 1999). However, be careful not to make clients feel as if they are being abandoned. To the extent that you can, try to maintain some flexibility in your schedule to accommodate the time requirements of an extremely complex and/or difficult case.

- Summarize the session. You can do this in several ways. First, you can provide an overview of the discussion and ask your client how s/he feels about what you've discussed. Another option is to ask your client to provide a summary (e.g., "Our time is about up. I was wondering what stands out for you as far as what we've covered today"). You should fill in missing information and/or correct inaccurate patient statements; frequently you will need to correct technical/factual information. Next, you might ask your clients to briefly describe where they are in their decision-making process. For example, "What are your thoughts and feelings about the next thing you will need to do in order to make a decision?" Finally, if appropriate, discuss how abnormal results will be conveyed to them, and possibly walk clients through a discussion of *what if* the results are abnormal.

- Discuss next steps. Review what will happen next and what actions they can take. For example, "In about 10 days you will receive a letter summarizing what we discussed today. I will call you in two weeks with the results of the amnio. If you have not heard from me by then, please call the center. Also, if any other questions or concerns come up, please feel free to call me here."

- Arrange for follow-up. If you will interact with the client again, explain how future contacts can be made. Schedule an appointment now, if feasible. Also, keep the door open, letting clients know that they may return at some future time if they need to do so (Kramer, 1990). But do this carefully, since this may be a way to avoid truly ending the relationship. Furthermore, clients may not have health care coverage for additional sessions.

- Try to end on a positive note, if appropriate. Remember, not all clients will hear good news, so be careful not to offer false hope and reassurances. Clients who receive bad news will experience any number of negative feelings (anger, grief, anxiety, shock, despair—see Chapter 9 for a discussion of client affect). You may be tempted to try to make them feel better. This is probably not possible nor even desirable at this point. Be careful not to offer platitudes such as, "Everything will be fine"; "Things will look better after a good night's sleep"; "You'll get over this in time"; "It's all for the best." In Chapter 7, we discuss communicating bad news to clients.

- Reinforce clients. Express confidence in their decision-making processes and in their ability to get through this difficult time. Be careful to reinforce their process and not their actual choice! For example, "You have done some very careful thinking about your options and seem to know which one is best for you" rather than "You're doing the right thing by continuing this pregnancy."

- Assume responsibility for ending. You will have to be assertive, especially with clients who are reluctant to end (Geldard, 1989). For instance, in a prenatal center clients often are scheduled hourly, and you can't go over time. When you realize that you are going long, halfway through the session say, "I realize we only have __ minutes left. So I want to be sure we've covered the most important things for you." Also, as we mentioned previously, at the beginning of the session tell clients how much time you have, and place a clock where clients can see it. In some extreme situations, you may have to say, "I realize that you'd like to say more, but we need to finish now." Look at your watch, stand up, move toward the door, and open it. These cues will signal that the session is over.

- Don't invite additional conversation. If you are trying to wind down, avoid making a reflection of client content or affect because it encourages clients to elaborate. Avoid asking questions for the same reason (Geldard, 1989). Instead, consider these strategies for ending: change the subject; slow your pace of talk; you talk more, and the client, by default, talks less.

- Be sensitive to client emotional state. If clients are crying and/or otherwise visibly distraught, allow them a few minutes to compose themselves before leaving the room.

- Say good-bye. Let clients know that you were glad to have an opportunity to work with them and that you wish them the best. Sharing your feelings about the end of a session or genetic counseling relationship models for the client an appropriate way of closing the contact (Lanning and Carey, 1987).

- Observe social amenities regarding departure, hold the door for clients, escort and/or direct your clients to the exit; and shake hands, but only at their initiation.

- Don't counsel outside of the room. If you have to escort your clients any distance to an exit, some may attempt to continue counseling with you. Try to direct the conversation away from genetic counseling to social conversation about the weather or about where they had to park their car, etc.

- Allow yourself enough time to debrief about the ending process. Kramer (1990) eloquently describes the importance of psychotherapists' allowing themselves enough time, "It struck me when I was terminating with a long-term client and not allowing myself a few minutes before the next patient, that I was not doing the best thing either for the terminating client, for myself, or for my next patient. This kind of scheduling does not give the relationship the respect it deserves and thus diminishes its import....By not allowing myself those minutes I did not give myself the time to mourn, celebrate, or feel what I needed to feel" (p. 72). This same recommendation seems appropriate for genetic counseling, especially when you are ending a genetic counseling relationship of some duration and/or intensity. We suggest that you discuss your reactions about ending with your clinical rotation supervisor (both before and after the final session).

• Respect a client's autonomy to end early (Kramer, 1990). Some clients may want to end before you feel that everything has been adequately covered. There can be several reasons why clients might want to leave prematurely, including client discomfort with difficult and/or painful information; client denial that anything is wrong; you give the impression that you are overwhelmed with the intensity of their situation or reaction; and you fail to adequately assess the complexity of the client's situation (both medical and psychosocial) and/or the client's response to it. You need to respect the client's wishes to end early. However, remember that there is some information that you must present, such as risk. One option is to send a follow-up letter detailing the information that you feel requires more explanation.

Difficult Genetic Counseling Endings

There are a number of situations in which ending the genetic counseling relationship may be difficult. Generally speaking, the longer you've worked with clients, or the greater number of contacts you've had with them, the more difficult the endings (Pinkerton and Rockwell, 1990). Some clients may feel dependent on you, or they may have really enjoyed working with you (both are more likely if you've had more than one interaction). Research from the mental health field also suggests that endings are more difficult when the process and outcome have not gone well; in other words both the client and the counselor are dissatisfied (Brady et al, 1996; Quintana, 1993). This difficulty is probably due to the fact that the counselor and client have not been able to establish a comfortable working relationship (Quintana, 1993). For instance, you may question whether you could have done more; you may feel guilty about not having been more effective; your view of yourself as a skilled genetic counselor may be challenged by an angry, dissatisfied client who leaves early; or the client may not have adequately integrated information from the genetic counseling sessions. In addition to relationship endings, it can be challenging to end individual genetic counseling sessions, especially with highly verbose clients, emotionally distraught clients, clients who are making what you consider to be the wrong decision, clients to whom you've given bad news, and clients with whom you feel a strong connection (e.g., you wonder how things will turnout for them).

Perhaps one of the most challenging endings is with terminally ill clients where your good-byes are symbolic of their eventual deaths (e.g., a 17-year-old boy with Duchenne muscular dystrophy). Sometimes counselor difficulties with endings are related to unresolved endings in their own lives. Ending a genetic counseling relationship may represent these unresolved issues (see Chapter 12 on transference and countertransference). Clues that you are having trouble with endings include consistently exceeding session time limits, and looking for excuses to recontact clients.

Try to anticipate particularly difficult endings whenever possible and carefully plan for them. For example, you could discuss your feelings with your supervisor and brainstorm about how you might best proceed with saying good-bye. You could work on coming to terms with or accepting the limitations of genetic counseling, that is, it may not be the solution to every client's problems. We also recommend that you work on being aware of countertransference (unconscious reactions based on experiences from your own life; see Chapter 12). Try not to let your personal feelings interfere with the ending (e.g., unreasonable fears about what the client will do, your own feelings of not being helpful, etc.).

Making Referrals

Some of your genetic counseling clients may benefit from referrals to other sources of information, treatment, guidance, and/or support. Typically you would offer a referral at the end of the session or genetic counseling relationship, as part of the termination process. Reasons for referral include client issues that are beyond your expertise (e.g., client is suicidal and in need of psychiatric care; or client needs medical management for his or her condition); you do not have sufficient time to deal with the client's concerns (e.g., intense marital conflict precipitated by the confirmation of a genetic condition); or the client needs a broad network of support services (e.g., families who have a child with Fragile X syndrome) (Downing, 1985; Weinrach, 1984).

Making an effective referral requires careful planning on your part. It is your responsibility to first assess whether your clients might benefit from referral to other sources of help, and if so, to explain how the referral is intended to be in their best interests (e.g., explain why you are suggesting additional support/help and what they might gain from using additional resources) (Downing, 1985; Fine and Glasser, 1996). We recommend the following referral guidelines:

Building a Referral Base

- Be familiar with referral sources. You should build and continually update a referral file that includes the names, addresses, telephone numbers, and procedures for contacting various referral sources (Cheston, 1991; Fine and Glasser, 1996). Build a file by asking colleagues for recommendations, by checking with clients for sources that have been helpful to them, and by familiarizing yourself with local social services (Fine and Glasser, 1996). You should update your file periodically (e.g., checking on which referral sources are currently accepting new clientele) (Cheston, 1991).

- If you live in a small, isolated locale, pool your resources with all of the health care professionals in your area in order to establish a limited referral group, and broaden the group with whom you pool resources to include nurses, physicians, religious leaders, psychologists, social workers, teachers, etc. (Cheston, 1991).

- In choosing referral sources for your file, select sources that are sensitive to and aware of cross-cultural issues and gender issues, are located a reasonable distance from clients, and are affordable (Cheston, 1991).

- Foster good relationships with the people in your referral file so that your clients will get prompt attention from the appropriate individuals (Fine and Glasser, 1996).

Points to Consider When Making Referrals

- Consider your client's resources before making a referral. For example, be aware of expense, distance to the source, and whether the client has insurance or some other means of paying for the referral source (Weinrach, 1984).

- Offer clients more than one resource if possible (Cheston, 1991; Downing, 1985) so that they have some choice and do not get the impression that you have a vested interest in a particular source.

- Avoid making hasty, inappropriate referrals because you feel unsure of you own ability to be helpful, you are trying to slough off a highly disruptive client, or you think so highly of a referral source that you refer all of your clients there regardless of the appropriateness (Weinrach, 1984).

- Referrals do not always have to be made immediately. For instance, services may not be available, or the client may not be emotionally ready. Let your client know that your referral sources are available in the future.

- Refer to specific individuals within clinics or agencies, if possible (Downing, 1985).

- Offer the referral tactfully so that clients feel it is to provide them with maximum assistance and not that their problems are so severe that they need extra help (Downing, 1985). Start by focusing on the importance of their problems or needs and the desirability of resolving the problem.

- Prepare your client. Provide details about the referral source (e.g., name, location, fees) to lessen anxieties about this new relationship. Describe the competencies and characteristics of the referral person(s), and how to contact the referral source. You are trying to enhance this person's credibility and expertise.

• Don't be too prescriptive with respect to the type of service or treatment the client will receive from the referral source. For instance, do not suggest that your client would receive a certain type of medication (Downing, 1985).

• Cheston (1991) cautions against overselling the referral (i.e., be careful to give an honest description of the service rather than promising that it is the perfect source for your client).

• Coordinate the referral. In some cases you will want to contact the source to be sure that it will take the referral. But keep your client's identity confidential until s/he decides to take the referral (Cheston, 1991).

• Use a written referral form because clients are more likely to follow up on written referrals as opposed to verbal ones (Cheston, 1991). This would be particularly important for distraught clients and/or for clients who do not have English as their first language; such clients might have difficulty comprehending and retaining verbal information. Indicate on the written form all of the referral sources being offered, addresses and telephone numbers, and consider drawing a map on the back to assist clients in finding the referral sources (Cheston, 1991).

• Check out client feelings about the referral. Even clients who seem to readily accept it may be apprehensive. Normalize any fear. For instance, "Most people would feel hesitant to bring this issue to their minister" (Cheston, 1991). You could point out how the client took the risk to come to see you, and build on this to get the client to take the referral (Cheston, 1991). In some situations, you may decide to call and make the appointment, with the client's permission.

• Anxious clients may ask a lot of questions about the referral source, and you should patiently answer their questions (Cheston, 1991). In any case, ask your clients what they feel about the referral, and then what they think about it (Cheston, 1991).

• Clients who get angry about a referral may feel rejected or believe they are being passed off to yet another agency or professional, or be scared about the severity of their concerns (Cheston, 1991). For example:

Client: "So, you're just blowing me off, too!"

Counselor: "You think I'm not being helpful?"

Cl: "Yeah, you're just trying to get rid of me."

Co: "I think that you have some important questions that I cannot answer. I am referring you to the best agency in this area because I want you to get the information you need to make your decisions."

• In some situations, you may want to provide a written summary of the genetic counseling relationship to the referral professional (Cheston, 1991).

For example, if a client will see a psychotherapist, the information from the genetic counseling session would enhance her care. Be sure to get written client permission before sending a report.

- Include the referral in a follow-up letter in order to remind clients.

- For some clients (e.g., a child referred to a special education professional), you may wish to follow up to see if the referral was taken, and perhaps schedule another session with the client after s/he has met with the referral source. Alternatively, you might ask the client to call you in a couple of weeks to let you know how things went (Downing, 1985).

- Remain available to answer any questions your client might have about the referral source (Weinrach, 1984).

- Cheston (1991) cautions against leaving clients to their own resources. If your client is asking for additional help, and you are not immediately aware of any that you can provide, then tell your client you will look into possibilities after the session and get back to her or him with whatever you find. Then follow through on your promise.

Table 6.2 contains a list of the types of referral sources that might be appropriate for genetic counseling clients.

TABLE 6.2. Common Referral Sources.

Support groups (genetic condition)
Bereavement groups
Agencies (SSI, medical assistance, WIC, social services, respite care)
Services (infant stimulation, schools, Association for Retarded Citizens)
Medical (genetics as well as other specialists)
Psychological (short-term therapy, long-term therapy, psychiatric, vocational counseling, family therapy)
Financial services
Adoption agencies
Social workers
Internet addresses
Clergy/spiritual leaders
Parenting classes
Drug/alcohol rehabilitation
Domestic abuse centers
Homeless shelters
Food banks
Developmental specialists
Unemployment center
Parents/individuals who are experienced with a condition and who are willing to talk with recently diagnosed clients/families

Closing Comments

You have a great deal of responsibility for beginning and ending genetic counseling sessions and relationships and for helping clients establish feasible goals. Advanced preparation, observing common courtesies, carefully listening to your clients, and observing time limits at the beginning and the end of sessions will assist you in these important genetic counseling activities. These responsibilities will become less challenging for you over time and with experience. You will gradually develop your typical way of structuring genetic counseling relationships and sessions.

Class Activities

Activity 1: Discussion:
Think-Pair-Share Dyads and/or Small Groups

Students discuss the following:

- What concerns/questions do you think that clients have about genetic counseling?
- What client questions are you unsure about how to address? Afraid to address?
- How much welcoming should you do? Should you engage in chit-chat?
- Should you go to the waiting room to get your client? Do you escort your client to the exit at the end of the session?
- How much structure should you give your clients about the genetic counseling process?
- Should you provide follow-up, and if so, how?

PROCESS

The whole group discusses their responses to these questions.

Estimated time: 20 minutes.

Activity 2: Model

Instructor and a volunteer genetic counseling client engage in a role-play in which the counselor demonstrates how to initiate the genetic counseling session. Students observe and take notes of counselor behaviors.

Estimated time: 10 minutes.

PROCESS

Students discuss their examples of counselor initiating behaviors. They also discuss the effect that these genetic counselor behaviors appeared to have on the client.

Estimated time: 10 minutes.

Activity 3: Triads

Students practice initiating genetic counseling sessions using the three role-plays described below. The client selects one of the following client roles without letting the counselor know about the role in advance.

CLIENT ROLES

1. You are here to see a genetic counselor because you are at risk for Huntington disease. You are worried that potential employers will find out that you have the gene. You also worry about how you will react if you find out that you have the gene. You feel that, at some level, suicide might not be out of the question.

2. You are here to see the genetic counselor because of advanced maternal age. You have had some very bad experiences with health care professionals in the past, and feel as if you have been railroaded into coming for genetic counseling. You question what the genetic counselor sitting in front of you will be like. What are her/his credentials? How can you be sure that s/he will be helpful?

3. You are here to see the genetic counselor because of an abnormal triple screen. You are very scared that something could be wrong with your baby. You have been told by members of your cultural group that genetic counselors can give your baby problems. Furthermore, you are afraid that if you go through with the genetic counseling, if will mean that you have lost faith in God's being able to handle things. Still, you desperately want to know if your baby is OK.

Estimated time: 75 minutes.

PROCESS

The whole group discusses: What did you learn from these role-plays? What interpersonal skills were required?

Estimated time: 10 minutes.

INSTRUCTOR NOTE

• Observers can use Appendix 6.2, Observer Checklist for Genetic Counseling Interview.

Activity 4: Think-Pair-Share Dyads and/or Small Groups

Students discuss client and counselor resistance to ending. What might be some reasons that clients would not want to end the genetic counseling session and/or relationship? Do you possess any characteristics that would make it difficult for you to observe time limits? To end the session and/or relationship?

The whole group discusses their responses to these questions.

Estimated time: 25 minutes.

Activity 5: Triads Exercise

Students practice ending a genetic counseling session using a combination of the genetic counselor and client roles described below. The counselor and client should not reveal which role they chose to portray until the end of the role-play. They should spend 10 minutes doing the role-plays and about 10 minutes after each role-play to provide feedback.

Genetic Counselor	Client
Resist ending	Resist ending
Encourage ending	Encourage ending
Uncertain about ending	Uncertain about ending

Estimated time: 60 to 75 minutes.

The whole group discusses: What did you learn from these role-plays? What interpersonal skills were required?

Estimated time: 10 minutes.

- Observers can use Appendix 6.3, Observer Checklist for Counselor Ending Genetic Counseling.

Activity 6: Think-Pair-Share Dyads and/or Small Groups

Discuss referral. What do you think is particularly difficult about making effective referrals? (Consider both client and counselor characteristics as well as external factors such as the referral sources themselves.)

The whole group discusses their responses to these questions.

Estimated time: 15 to 20 minutes.

Activity 7: Triads: Making a Referral Role-Play

Students engage in 10-minute role-plays. The client is a prenatal genetic counseling client who the counselor wants to refer to a psychotherapist. The reason for the referral is that the genetic counselor believes that she has unresolved grief over prior multiple pregnancy losses. The genetic counselor realizes that these issues are too complex to address during the genetic counseling session. In the first role-play, have the client accept the referral. In the second role-play, have the client resist the referral. In the third role-play, repeat either of these two client roles. Follow each role-play with 10 minutes of feedback.

Estimated time: 60 minutes.

Process

The whole group discusses: What did you learn from these role-plays? What interpersonal skills were required?

Estimated time: 10 minutes.

Instructor Note

• Observers can use Appendix 6.4, Referral Checklist.

Activity 8: Think-Pair-Share Dyads and/or Small Groups

Discuss these questions: What are your reactions to setting goals in genetic counseling? What might be some difficulties in setting goals from the client's perspective? From your perspective? How do you avoid imposing your goals on clients?

Estimated time: 10 to 15 minutes.

Activity 9: Brainstorming: Unrealistic Genetic Counseling Goals

Small groups generate a list of unrealistic genetic counseling goals for clients.

Estimated time: 15 minutes.

Instructor Note

• This activity could be expanded by having dyads or small groups rewrite the goals to make them more realistic.

Activity 10: Goal Setting

Dyads or small groups turn the following genetic counseling client problem statements into goal statements:

• A prenatal patient says, "I'm afraid something's wrong with my baby."
• "My son has just been diagnosed with Klinefelter syndrome, and he needs someone to talk to."

- "My mother died of breast cancer, and I'm afraid I'm going to die like that, too."
- "I'm sick of wondering whether or not I have the gene for Huntington's!"
- "No one can tell me what's wrong with my son."
- "I was hoping there was some research that could help my daughter with this disease."
- "I don't want to have any test that's going to hurt my baby."
- "My doctor sent me because I'm pregnant and over 35."

Estimated time: 15 minutes.

INSTRUCTOR NOTE

- This activity could be done individually as a written exercise.

Activity 11: Model

The instructor and a volunteer genetic counseling client engage in a genetic counseling role-play in which the instructor demonstrates how to establish goals with the client. Students observe and make notes about counselor behaviors.

Estimated time: 15 minutes.

PROCESS

Students discuss their examples of counselor goal-setting behaviors. They also discuss what goals were established and give their impressions of the impact of the goal-setting process on the client.

Estimated time: 10 minutes.

Activity 12: Triads: Goal-Setting Role-Play

Students use the following client roles to engage in 15-minute role-plays in which they attempt to establish genetic counseling goals with their clients. Follow each role-play with 10 minutes of feedback.

CLIENT ROLES

1. Advanced maternal age client coming to clinic for prenatal genetic counseling.
2. Twenty-five-year-old whose father was recently diagnosed with Huntington's disease.
3. A prenatal client who has a child with cystic fibrosis.

Estimated time: 75 to 85 minutes.

PROCESS

The whole group discusses: What did you learn from these role-plays? What interpersonal skills were required?

Estimated time: 10 minutes.

Written Exercises

Exercise 1

Engage in a 20-minute role-play of a genetic counseling session with a volunteer from outside of class. During the role-play, focus on beginning the genetic counseling session and setting goals. Audiotape the role-play. Next transcribe the role-play and critique your work. Give the tape, transcript, and self-critique to the instructor who will provide feedback. Use the following method for transcribing the session:

Counselor	Client	Self-Critique	Instructor

Exercise 2

Engage in a 15-minute role-play of a genetic counseling session with a volunteer from outside of class. During the role-play, focus on ending the genetic counseling session and making a referral. Audiotape the role-play. Next transcribe the role-play and critique your work. Give the tape, transcript, and self-critique to the instructor who will provide feedback. Use the following method for transcribing the session:

Counselor	Client	Self-Critique	Instructor

Exercise 3

Develop a referral file using the categories listed in the Table 6.2. Include names, numbers, addresses, types of services provided, and fees.

INSTRUCTOR NOTE

- This could be done as a group project.
- Students can be instructed to build a list of national resources as well as local ones. The national resources would be useful to them regardless of where they end up practicing.

Exercise 4

Develop an interview checklist that you could actually use in your genetic counseling sessions (see Appendix 6.1 for examples of possible topics).

INSTRUCTOR NOTE

- This can be done as an in-class activity, where small groups develop a list for different case scenarios such as elevated maternal serum alpha-fetoprotein (MSAFP), newborn with Down syndrome, advanced maternal age, or known genetic condition in the family.

Exercise 5: Client Goals*

For each client scenario below, write possible client genetic counseling goals:

- A couple is referred for genetic counseling for consanguinity (first cousin union). They are from Italy. They have not told their families about their relationship. They are in their early 20s.
- The clients are a young, Hispanic couple who have brought their 2-year-old son to see the medical geneticist because of developmental delay. They are uncertain as to why they need to see the medical geneticist, but are doing what their pediatrician recommended.
- The client is a 31-year-old white, g5, P0040 (5th pregnancy, 4 abortions— not elective) woman who is pregnant for the fifth time. She has had four miscarriages, and she has no living children. She has a history of seizures and multiple pregnancy loss. You are seeing her in this pregnancy to discuss the possible teratogenic effects of her seizure medications. (Her seizures have occurred only in the past 1 1/2 years due to a serious car accident.) She has a normal triple screen and targeted ultrasound. Later, severe intrauterine growth retardation (IUGR) is noted at 25 weeks. An amnio is done and results reveal trisomy 21, Down syndrome. The patient is distraught that she may lose another baby and is also dealing with an unhappy marriage and home life. To make matters worse, she has terrible short-term memory due to the accident, and every time you see her, you have to repeat most of the information.
- The client is a 24-year-old white woman with a history of epilepsy. She is planning a pregnancy next year.
- The couple have had 7 years of infertility and have learned they are pregnant with twins. The mother is 39 years old.
- The client is a 6-year-old boy who is developmentally delayed. The family has been to numerous doctors, and someone suggested that he be tested for Fragile X syndrome.

*Adapted from Cormier and Hackney, 1987.

- The client is a 37-year-old, g2, P1001 (second pregnancy, 1 full term delivery, 1 living child), woman pregnant for the second time. She is approximately 32 weeks pregnant. She has had one normal delivery resulting in one normal living child. Amniocentesis revealed trisomy 7 mosaicism in cultured cells. Ultrasound and fetal echo were normal.

- The client is a 14-month-old boy with severe encephalopathy. The family history is significant for a similarly affected male sibling who died at 18 months of age. All studies (done elsewhere) have been negative; no diagnosis has been provided to the family, yet they have scheduled an appointment in your clinic.

- The client is a 13-year-old girl who is suspected of having Turner syndrome. The pediatrician who referred her spoke only to her parents about his suspicions. Although the parents have researched Turner syndrome on the Internet, they have told the client nothing about their concerns or about the condition itself. The client seems nervous and somewhat frightened.

Exercise 6: Obstacles to Client Goals*

For each of the client scenarios in Exercise 5, identify possible roadblocks (lack of knowledge, lack of skill, fear of risk-taking, and lack of social support) and generate interventions to address one roadblock from each category (i.e., knowledge, skill, risk-taking, and social-support).

Annotated Bibliography

Brown, D. (1997). Implications of cultural values for cross-cultural consultation with families. Journal of Counseling and Development, 76, 29-35.
[Discusses cultural empathy and counselor sensitivity to cultural differences, and describes several ways that different cultural groups may vary in their approaches to goal-setting and problem-solving.]
Cheston, S. E. (1991). Making effective referrals. New York: Gardner Press.
[Written for mental health professionals, the book contains useful information about the purpose of referrals, roadblocks to referring clients, and offers a number of practical suggestions for making referrals.]
Danish, S. J., D'Augelli, A. R. (1983). Helping skills II: Life-development interventions. New York: Human Sciences Press.
[Provides a detailed outline/guide for transforming client problems into positive goal statements, identifying roadblocks to goal attainment, and strategies for removing the roadblocks.]
Downing, C. J. (1985). Referrals that work. School Counselor, 32, 242-246.
[Although written for school counselors, the article contains several practical tips and suggestions for making referrals.]
Hackney, H. L., Cormier, L. S. (1996). The professional counselor: a process guide to helping (3rd ed.). Needham Heights, MA: Allyn and Bacon.

*Adapted from Danish and D'Augelli, 1983.

[Contains a chapter that describes how to establish goals in mental health counseling. The authors provide a number of useful guidelines.]

Rodolfa, E. R., Hungerford, L. (1982). Self-help groups: a referral resource for professional therapists. Professional Psychology, 13, 345-353.

[Describes the characteristics of self-help groups and their possible advantages and disadvantages.]

Taussig, I. M. (1987). Comparative responses of Mexican Americans and Anglo Americans to early goal-setting in public mental health clinics. Journal of Counseling Psychology, 34, 214-217.

[Reports data suggesting that clients of varying racial-ethnic backgrounds have positive reactions to goal-setting in mental health relationships.]

Vriend, J., Kottler, J. A. (1980). Initial interview checklist increases counsellor effectiveness. Canadian Counsellor, 14, 153-155.

[Provides a rationale for using a checklist within mental health counseling. Includes a list of possible checklist topics.]

Wilkinson, G. (1989). Referrals from general practitioners to psychiatrists and paramedical mental health professionals. British Journal of Psychiatry, 154, 72-76.

[Discusses some of the reasons for referral of medical patients to psychiatric/mental health services.]

Appendix 6.1: Suggested Topics for a Genetic Counseling Session Checklist*

Topic	Yes	No
Introduce self		
Explain genetic counseling process	____	____
Assess clients' reasons for seeking genetic counseling and what they expect from the session	____	____
Establish genetic counseling goals	____	____
Gather family history	____	____
Explain natural history of condition	____	____
Explain inheritance	____	____
Discuss risk rates	____	____
Present possible options	____	____
Explain genetic test(s)	____	____
Observe clients' mood/counseling issues	____	____
Explore clients' thoughts and feelings	____	____
Summarize session	____	____
Make referral	____	____
Explain follow-up procedures	____	____
Close the session	____	____

*Adapted from Vriend and Kottler, 1980.

Appendix 6.2: Observer Checklist for Genetic Counseling Interview

	Yes	No
Initial greeting		
Chairs at comfortable distance	____	____
Faces client	____	____
Introduces self	____	____
Asks how to address client	____	____
Makes some small talk, if appropriate	____	____
Informed consent		
Presents credentials	____	____
Discusses taping/supervision	____	____
Discusses confidentiality limits	____	____
Describes genetic counseling	____	____
Problem exploration		
Inquires about clients' reasons for seeking genetic counseling	____	____
Attempts to establish goals	____	____
Seeks clients' agreement to work on goals	____	____
Miscellaneous		
Is sensitive to clients' questions/concerns	____	____
Discusses follow-up to the session	____	____
Additional comments		

Appendix 6.3: Observer Checklist for Counselor Ending Genetic Counseling

	Yes	No	N/A
Anticipate conclusion of session			
Summarizes major points	____	____	____
Reaches consensus on what to do next	____	____	____
Agrees on time, date, and place of future contact	____	____	____
Observes session limits	____	____	____
Gives cues to signal the end	____	____	____
Determines why client does not want to leave	____	____	____
Makes special arrangements for additional time	____	____	____
Observes courtesies of departure	____	____	____
Ends on positive note, if appropriate	____	____	____
Structures end of relationship			

Assesses extent to which goals were accomplished	____	____	____
Asks client to summarize decision-making progress	____	____	____
Summarizes client decision-making progress	____	____	____
Plans follow-up	____	____	____
Makes referral to other sources	____	____	____
Additional comments			

Appendix 6.4: Referral Checklist

	Yes	No
Describes clients' issue/problem for which the referral is made	____	____
Selects referral sources appropriate for the clients and their situation	____	____
If possible, provides more than one name for referral	____	____
Informs clients about referral source name, address, phone number, and how to contact	____	____
Contacts the referral source directly	____	____
Presents referral source credentials to enhance credibility	____	____
Makes positive statements about the usefulness of the source for helping the clients	____	____
Secures clients' written permission to send a report of the genetic counseling to this source	____	____
Follows up with referral source to see if referral was taken	____	____
Follows up with clients to inquire about their experience with the referral source	____	____

CHAPTER 7
_____

Collaborating with Clients: Providing Information and Assisting in Client Decision Making

---

Learning Objectives

1. Define the skill of information giving.
2. Identify strategies for presenting risk information.
3. Differentiate giving normal and abnormal test results.
4. Identify factors affecting client decision making.
5. Develop skills for facilitating client decision making through practice and feedback.

Two major genetic counselor activities are providing information and assisting clients in their decision making. Some genetic counseling sessions are primarily informational or supportive in nature. For instance, parents of a newborn with Down syndrome may need support to deal with what is present as opposed to dealing with decision making. Other sessions involve clients who are facing one or more decisions, and providing information is an essential intervention. The information that you present is an integral part of collaborating with clients, helping them understand and sort through factors relevant to their decision.

Providing Information

Information giving is a skill in which you provide facts, details, and data in an attempt to help clients be as fully informed as possible about their risks, related medical information, possible options, and potential outcomes. "Genetic counselors have a daunting task. They have to transmit an enormous amount of complex information to people who know little about genetics, and they have to explain words that may have a variety of meanings for different people, even within the medical profession" (Chapple et al, p. 87). In many situations you will have to provide complicated and ambiguous information such as risk rates, sensitivity and specificity of tests, and severity of conditions diagnosed prenatally such as Down syndrome (McCarthy Veach et al, 2001). It is your responsibility to present this information in a comprehensible way.

Information giving differs from advice, in that advice is an attempt to suggest what your client should do. Information giving, when done effectively, provides clients with knowledge that can help them choose their own course of action. There are two ways that you can present information to clients. One way is to respond to client questions. Clients vary in the extent to which they ask questions, although you may be able to invite questions by asking, "What questions do you have?" A second way to provide information is to determine what additional facts or details may be pertinent to a given client's situation and to present them. In either case, you should keep in mind that clients vary in the extent to which they wish to hear information, especially scary and discouraging information.

The types of information that you present during genetic counseling sessions can be broadly categorized as (1) information that sets the agenda and describes what to expect during the session (see Chapter 6); (2) information pertinent to clients' reasons for seeking genetic counseling (e.g., risk rates, testing options, disease information, support services); (3) information that facilitates clients' decision-making processes (e.g., possible options, outcomes); and (4) follow-up information about what will happen after the session.

Providing Risk Information

There are no clear directions about how to ensure that the probabilistic nature of risk estimates is accurately communicated and understood. Additionally, there is uncertainty about how to sensitively communicate the error-proneness of genetic tests. [Bottorff et al, 1998, p. 77]

Risk information is one of the most complicated types of data you can present during genetic counseling. The term risk is defined differently by different individuals, and people vary markedly in their perceptions of the relevance and meaning of risk data for themselves and others (Bottorff et al, 1998; Hallowell et al, 1997; Palmer and Sainfort, 1993).

So how can you best present risk information? As percentages? Numbers? With verbal descriptions? Does your client have a 1 in 100 chance of having a genetic condition, or a 99 in 100 chance of not having it?

Hallowell et al (1997) found that there is a great deal of variability in how genetic counselors present risk information. They also found that a majority of their sample of genetic counseling clients who received risk information concerning breast and ovarian cancer preferred quantitative data because these types of data provide clearer and more concrete information. The researchers found little difference in client preference for percentages, proportions, or population comparisons. Despite their findings, you should remember that not all clients prefer and/or understand presentations that are purely quantitative.

For example, Green and Murton (1993) found a fairly high level of dissatisfaction with numerical risk estimates by mothers who received genetic counseling for Duchenne muscular dystrophy. Additionally, clients may have different preferences for receiving information about different types of risks, for example, reproductive risks versus health risks (Hallowell et al, 1997).

Nevertheless, these findings suggest that it is generally a good idea to present some quantitative data. Additionally, because some clients prefer qualitative descriptions (e.g., the risk is high or low, or likely or unlikely), we recommend that you include both quantitative and qualitative presentations. You generally should avoid providing only qualitative data since they can be misinterpreted by the client and/or reflect your biases (Sagi et al, 1998).

There are five types of quantitative formats for providing risk information (Hallowell et al, 1997):

- proportions
- percentages
- ratios
- odds against
- as comparisons with population risks

The most effective approach may be to present individual risk as a comparison with population risks since it can help clients put their personal risk in perspective (Hallowell et al, 1997).

Client Factors Relevant to Risk Perception

A number of client factors may affect how they perceive their risk. As described by Bottorff et al (1998), these include:

- Cognitive functioning: for example, the extent to which the client thinks abstractly and mathematically. (This is related to intelligence.)
- Temperament and personality: for example, pessimists may inflate risk figures while optimists underestimate their risks; and achievement-oriented individuals may believe they can beat the odds, while failure-threatened individuals believe that there is little to no chance of winning. "When there is a moderate to high risk of recurrence of a condition, the person who is emotionally positive may think that the 1 in 4 risk is 'good news' because the odds are in favor (3 in 4) of a healthy child being born. On the other hand, a person who is emotionally depressed may see a 1 in 20 or 5% risk of recurrence (a low genetic risk) as being unacceptable because they may feel 'everything happens to me'" (Murray, 1976, pp. 14-15).
- Attributions/worldview: for example, individuals may have an external locus of control such that they attribute outcomes to chance or fate.
- Personal experience with a condition: individuals with a family history of a condition do not view risk estimates hypothetically.

- Degree of understanding of the general population risk.

- Coping styles: for example, information seekers, avoidant style, dependent decision makers, minimizers, etc.

- Gender: males and females may perceive the same information differently.

- Temporal factors: for example, time changes how people view their situation.

- Cultural or ethnic identity: for example, in a study of Hmong refugees' English proficiency, Ostergren (1991) had to modify a rating scale of 0% to 100%, changing it to 1 to 3, because percentage was an unfamiliar concept.

- Religiosity: one's values, philosophy, meaning of life, etc., affect perceptions. For example, Somali immigrants are likely to believe that disability is caused by God (Greeson et al, 2001).

Simonoff (1998) includes these client characteristics as factors related to risk perception:

- Difficulties grasping the concept of probability.

- Unresolved issues such as not accepting the diagnosis in the proband; or believing another factor is causative—for example, recent Mexican immigrants may believe that the lunar eclipse causes birth defects (McCarthy Veach et al, 2001); or believing that falling during a pregnancy caused a child to have Down syndrome.

- Different perceptions by family members of risk, burden, and/or desire for more children. Simonoff (1998) offers this example about how the meaning of risk may vary: "Thus the family who believed all offspring would be affected may receive a 50% risk as good news. Similarly, and more relevant to autism, families burdened by a very disabled child may view a recurrence risk of 5% as unacceptably high" (p. 448).

General Guidelines for Presenting Risk Information

- Before presenting risk information, ask clients to describe preexisting perceptions of their risk (Bottorff et al, 1998; Sagi et al, 1998). For example, Sagi et al (1998) found that women in their sample tended to overestimate their risk of breast cancer.

- Your task is to translate extremely complicated technical information about risks into language that is comprehensible to your clients.

- In choosing when and how to communicate risk information, you should take into account your clients' desire for information and how much control they are attempting to exert over their decision-making process (Bottorff et al, 1998). One strategy is to ask clients, "What does the information mean for you?" or "Is this what you were expecting?"

- Try to remove your personal biases about risk from your presentation. For example, present both sides of a risk, "There is a 1% chance of ___ occurring, and there is a 99% chance of it not occurring." Terms such as, "the risk is only..." and "as high as..." suggest your own views of risk information; you should present risk information in a few different ways in order to present it more neutrally (Simonoff, 1998).

- You will probably have personal preferences for the way in which you present risk information. Indeed, Hallowell et al (1997) found this to be the case for their sample of experienced genetic counselors. Just remember that you may need to supplement your personal preference with another way in order to accommodate client preferences, abilities, situations, etc. In other words, you need to be flexible in how you communicate risk information.

- Avoid words that convey a stigma or negative connotation, such as "mutant," "defective," "abnormal" (Bottorff et al, 1998). Instead, you could say, "working and nonworking genes."

- Try to personalize the information to the client because people tend to remember information that is relevant to themselves (Sagi et al, 1998). For example, say something like, "This is *your risk*; this is what the outcomes might be *for you*," etc.

- It is important to remember that your client's perceptions of risk may be quite different from yours (Bottorff et al, 1998; Hallowell et al, 1997) and that many clients interpret risk as binary or categorical (e.g., either I have the gene or I don't have it) no matter how your present the data (Bottorff et al, 1998; Lippman-Hand and Fraser, 1979b).

- How you provide risk information has a significant impact on client interpretations of its meaning. "Informing an individual that he or she has a chance of a particular occurrence of 1.3 in 10,000 compared to the general population's chance of 1 in 10,000 is not particularly impressive to most people. However, if the format was such that the individual was informed that his or her risk is 30% greater than that of the average individual, the situation is likely to be seen as 'riskier,' although the two situations are equivalent" (Bottorff et al, 1998, p. 70).

- Emphasize that risk rates are probabilities, not guarantees. Try to help clients understand the uncertainties involved. This will be challenging because some clients regard genetic testing and information as prophetic (Bottorff et al, 1998). Additionally, some clients will perceive your discussion of uncertainty as evidence of your honesty, while others will interpret it as evidence that you are incompetent (Bottorff et al, 1998).

- Be aware that it is very difficult to communicate objective risk to clients. Bottorff et al (1998) offer an illuminating example: "Providing risk information about cancer to individuals who may perceive themselves to be healthy, and who may or may not have directly observed a close relative with the disease, requires these individuals to engage in sophisticated abstract thinking. This issue becomes more pronounced in situations where there is not effective therapy available or the information is only relevant in considerations of possible future outcomes" (p. 69).

- Remember that you may be unable to give precise risk percentages to some clients because available family history is limited; because existing empirical data on risks may be from samples that are not representative of your clients (Sagi et al, 1998); or because of other reasons such as testing limitations, and unavailable or equivocal data (e.g., some teratogens).

- Sagi et al (1998) point out that risk is not equivalent to probability. Risk perception actually involves both probability and adversity (or burden of an outcome). So one of your challenges is to assess how adverse a particular outcome would be for your client and to associate that adversity with the probability of the outcome's occurring. Try working with your client to identify all of the consequences of a particular outcome (medical, psychosocial, financial, etc.) and their relative importance to or likely impact on the client (Sagi et al, 1998).

- Ask clients to summarize their understanding of their risk after you have given them the information. This will allow you to correct any inaccuracies and will give you insight into their subjective perceptions of risk (e.g., "What is your understanding of the risk we've just discussed?").

- Be prepared to deal with client affect regarding personal risk. Emotions may include fear, anger, guilt, grief, shame, embarrassment, and lowered self-esteem (Bottorff et al, 1998). Try an empathy response in which you reflect your client's emotions. For example, "You have gotten very quiet since I gave you the information about your risk. Are you feeling scared right now?"

Common Mistakes When Giving Information

Because information giving is a skill, it requires practice and self-reflection to do it well. Common mistakes of beginning genetic counseling students include:

- Being disorganized. Prepare your information in advance. Gather relevant information and arrange it in a clear order, although you may not follow that order.

- Not understanding information yourself. Before meeting with your client, be sure that you comprehend complicated medical details, risk rates, etc., and have thought through how you will explain this information.

- Following a "cookbook" approach. The same recipe won't work for all individuals. You can't write out a script and follow it exactly with all clients. You should tailor and present the information to each client and her or his situation. This is a skill that you will develop over time and with supervised experience. You will find yourself gradually working more extemporaneously as you assess how a given client is responding to your presentation of information.

- Using technical terminology. Present information in language your clients can understand.

- Giving too much information. Initially, you may feel comfortable using elaborate drawings and providing extensive information. As you become more comfortable and gain experience, you will be better able to discern what is most relevant for each client.

- Giving clients information that they don't want. McCarthy Veach et al (2001) found that a major ethical and professional challenge is deciding whether or not to withhold information when clients don't want to hear it. For example, some clients may resent being given a detailed description of abortion when they have explicitly stated that abortion is not an option for them (McCarthy Veach et al, 1999). We suggest that you acknowledge a client's feelings about an option such as abortion. Then say that although you need to briefly present each option in order to fulfill your mandate to present all pertinent information, you will give the minimum amount of information on options that the client regards as unacceptable. For example, "I understand that you consider abortion to be unacceptable. I respect your feelings about that. I just wanted to make sure you were aware that it's an option, even if it isn't an option for you."

- Giving too much too fast. Proceed slowly so that your client can comprehend the information.

- Believing you have to present information in a certain order. Be flexible, providing information in a way that meets client needs.

- Failing to check on how much a client comprehends. Periodically ask your clients to summarize the information. For example, "What's your understanding of what we've just discussed?" Occasionally you may suspect that your client did not understand what you presented. In this case you might say, "Can you tell me what you've heard me say?"

- Saying, "Do you understand?" or "Do you have any questions?" Most clients will say that they understand, and many will deny having any questions. Instead, try open-ended statements: "What questions do you have?" and "What is your understanding of what I've said?" Then you should correct any inaccuracies.

- Getting into a lecturing mode, which lulls the client into passivity. Try engaging your clients by periodically asking for a summary or some other type of reaction during your presentation. Also try pausing occasionally during your presentation.

- Forgetting that cognitive content is emotionally charged. Your clients may have strong emotional reactions to factual information. Consequently, it will take them some time to digest and comprehend the facts. It is a good idea to repeat important information.

- Giving clients information without paying attention to their defenses and conscious and unconscious motivations (Mealy, 1984). You should try to recognize, for example, when clients are intellectualizing their situation and are making decisions without really understanding the long-range emotional impact these decisions will have (Mealy, 1984).

- Failing to accommodate cultural differences. Clients vary in their cultural interpretations of the causes and significance of genetic conditions (Punales-Morejon and Penchaszadeh, 1992). Try to determine their cultural beliefs and values and then frame your information within their cultural context.

- Thinking that you have to know all of the answers. Don't be afraid to say, "I don't know." Then find out, and get back to your client via phone and/or letter, or refer your client to another professional, if necessary.

Communicating Abnormal Test Results

One of the most difficult responsibilities that you will have as a genetic counselor is informing clients that their test results are abnormal, that they or their unborn child or family member has a genetic condition. This news has a tremendous impact on clients, and they may respond in a number of ways (crying, "shutting down," getting angry at you, etc.). Clients who test positive may feel an additional burden of worry and guilt about passing the gene to their children (Bottorff et al, 1998). This news may also have an intense emotional effect on you. For example, your empathy for clients makes you, to some extent, feel what they are feeling; you like your clients and don't want bad things to happen to them; you may feel like an executioner; you may feel guilty for being relieved that it's happening to them and not to you; etc. You have to allow yourself to experience these unpleasant emotions to a certain extent. Communicating abnormal results should never become routine. Your effectiveness as a genetic counselor depends in part on your ability to remain connected to your clients. At the same time, you need to balance this connection with a healthy distance from their situations. We recommend the following strategies for providing positive test results:

Prepare in Advance

- As we discuss later in this chapter, you could ask clients to engage in scenarios prior to undergoing testing. Ask them to imagine and describe what they will feel, think, and do if the results are abnormal. In addition to helping them anticipate their reaction, this strategy gives you some idea of how they might react when you present them with the test results.

- Practice giving abnormal results with a colleague so that you know what you want to say and how you want to say it.

- For a given client, visualize the meeting in which you present the results. Try to imagine in detail what each of you will say, feel, and do.

- Cultural variables can affect the way in which we provide information. For example, Jecker et al (1995) describe a case in which possible bad outcomes were communicated to a Navaho patient by referring to a hypothetical third party; this was done to reduce the patient's likelihood of thinking that he was "being witched." For some clients from Middle Eastern cultures, the family is a paramount resource for coping during times of crisis such as illness, and efforts may be made to shield vulnerable family members from unwelcome news (Lipson and Meleis, 1983). Therefore, a male family member (father or grandfather) may take the lead in the genetic counseling session. For Ethiopians, bad news generally should be told to a family member or close friend of the client who will then disclose this information to the client in a culturally appropriate way (Beyene, 1992). When you are unfamiliar with the cultural values and practices of your clients, you should do prior research on how abnormal test results are best communicated. For example, you could consult with community elders and review literature such as Fisher's (1996) *Cultural and Ethnic Diversity: A Guide for Genetics Professionals*.

Deliver the News

- Proceed slowly and calmly. Clients need time to absorb this type of news, and they will be more likely to express their feelings if you take a low-key approach.

- Consider whether you wish to deliver the news in stages (Faulkner et al, 1995). For instance: Counselor: "We have the results of your amnio, and they indicate a problem." Client: "What is it?" Co: "There's an unbalanced translocation." Cl: [Looks questioningly]. Co: "It's..."

- Allow clients to react the way they wish. Don't step in with false reassurances, don't encourage them to stop crying, don't go on at length with detailed information that they will not be able to hear. Information given too soon tends to block clients from expressing their feelings of disbelief, anger, grief, etc. (Faulkner et al, 1995). Sit quietly while they absorb the news, offer tissues after a few moments, and give them time to pull themselves together. You may want to pull your chair closer.

- Try to be yourself when communicating abnormal test results. Sometimes you will feel particularly connected to a client and will find yourself becoming tearful. It is all right to show some of your distress as long as it does not become the focus of the interaction. (Having tears in your eyes and feeling choked-up are probably OK; breaking down and sobbing are not.)

- Try saying, "I'm sorry." This simple phrase can communicate the depth of your feeling for your clients and their situation. This is an example of what Kessler (1999) describes as *providing consolation*: "A simple, sincere word or two or even a touch may have an enormous impact on persons who have been devastated by calamity" (p. 339).

Follow-Up

- Once your clients have regained some composure, ask whether they have any questions, then ask them to describe their understanding of what an abnormal test result means, and gently inquire about the next steps, that is, what the clients need.

- If relevant literature is available, you could give it to clients at the end of the session. This will allow them to take in the information at their own pace, time, and location.
- Referral to mental health counseling/psychotherapy may be appropriate if your client has an extreme emotional reaction.
- Talk with your supervisor to debrief after delivering abnormal test results. This will help you to relieve some of your emotion.

Communicating Normal Test Results

When you think about communicating normal test results your first impression might be that it is a simple and happy task. However, on closer consideration, you will find that it can be quite challenging. Some clients may not fully understand the meaning of a normal test result. For example, a normal amniocentesis does not mean that the baby is fine. Nevertheless, this is what some clients will think (McCarthy Veach et al, 2001). You must talk carefully and repeatedly about the limitations of a test. For example, "You didn't inherit the gene for breast cancer that your mother has. However, you still need to have regular checkups because you are at the same risk for breast cancer as any other woman"; or, "The baby's chromosomes are normal, but remember, that doesn't rule out all birth defects." Also, you must be sensitive to client frustrations that a diagnosis is not evident from testing.

Normal test results do not necessarily elicit only positive client emotions. Although many clients will feel relieved and happy, they also may have complex, mixed emotional responses that emerge over a period of time, and not all clients will feel relief at a normal test result. For instance, a client whose test for Huntington's disease is negative may feel guilty about her siblings who carry the gene and may be depressed about waiting as long as she did to have the testing done. Gray and her colleagues (2000) describe how one of the authors lived for years with the possibility of having the gene for Huntington's disease (HD). Once the individual finally pursued testing, which yielded negative results, this person's reactions were as follows:

I did not have Huntington's, but for the last 34 years I had lived in the shadow of the disease, and it took a long time to sink in that I was "normal"—I had the same chance as anyone else to live to be an alert, old person. I also grieved the time I spent not knowing and the choices I had made based on wondering if I had HD. I began to see the ways in which living with the threat of HD had limited my life choices and had narrowed my view of available options. More importantly, I felt that now I could have a life. The implications of this fact and my options were staggering! For a time, I was overwhelmed by the fact that I was HD-free and could live my life anyway I wanted, just like everyone else. [p. 9]

We suggest that you try to minimize any preconceived notions about how your clients will react and fully listen to their actual reactions. Similar to the suggestions above for abnormal test results, you can do the following:

- Ask your clients to consider different scenarios prior to undergoing testing. Ask them to imagine and describe what they will feel, think, and do if the results are normal. This will give you some idea of how they might react when you present them with the test results, although it is no guarantee that this is how they will respond.

- Allow your clients to react the way they wish. Sit quietly while they absorb the news.

- Reassure your clients that mixed feelings are a natural response, and try to draw out their emotions by asking them to tell you what they are feeling.

- Point out that their emotions may fluctuate and/or change over time, and inform them that you will be available to talk to and meet with them in the future when and if new concerns or questions arise.

Decision Making: Overview

Genetic counseling clients may be confronted with a number of decisions (e.g., whether or not to undergo testing, whether to continue or terminate a pregnancy when testing reveals a genetic condition, whether or not to have further/any children, who to tell and what to tell them, whether and when to undergo predictive testing, and whether or not to participate in related research). "A growing number of individuals and parents are confronted with genetic information. Not only does this increase their control and responsibility, it also forces them to make difficult and often painful decisions" (Huys et al, 1992, p. 17). Genetic counseling clients are under great stress: "They are vulnerable to many pressures and often are not willing or prepared to make a decision, even if given all of the information. For many, discussing personal issues such as contraception, finances, marital stability, or religion is an additional stress" (Mealy, 1984, p. 124). It is your role to assist clients with these difficult decisions.

Factors that Can Influence Client Decision Making

A number of factors can influence client decision making:

Client Decision-Making Styles

There are individual and cultural differences in decision-making styles. For example, Dinklage (1966) identified eight client decision-making styles:

- Intuitive: A person who makes a decision on the basis of feelings that have not been verbalized. This is the "it feels right" type of decider.

- Agonizing: A person who spends time and effort in gathering information and analyzing alternatives, but gets lost in the data and never gets to the decision point. This is the "I can't make up my mind" decider.

- Delaying: A person who puts off thought and action on a problem until later. This is the "I'll get back to you later" type. Unfortunately, this decider tends to never think through and make the decision. With this approach, deciding can be put off indefinitely.

- Impulsive: A person who takes the first available alternative, without looking at other alternatives or collecting pertinent information. This is the "decide now, think later" type.

- Fatalistic: A person who leaves the decision to the environment or to fate. This is the "whatever will be, will be" type. The individual believes that other forces outside of her or his personal control have tremendous influence. For example, strong religious beliefs, strong cultural beliefs in karma, etc.

- Compliant: A person who goes along with the plans of someone else rather than making his or her own decisions. This is the "if it's OK with you, it's OK with me" type.

- Paralytic: A person who accepts responsibility for a decision but is unable to set the process in motion to make the decision. This is the "I know I should, but I just can't get with it" type. The individual views decision making as too risky or frightening.

- Planful: A person who takes a rational approach to decision making, balancing cognitive, emotional, and situational considerations. This is the "I am planful and organized" type.

While all of these decision-making styles can be appropriate at certain times and in certain situations, the planful decision-making style may be useful when the stakes are high. In planful decisions, an individual moves through a seven-step process: (1) identifying the decision to be made; (2) gathering relevant information; (3) identifying alternatives or options; (4) weighing the evidence (e.g., will this option satisfy my need for support?); (5) choosing among alternatives; (6) taking action; and (7) reviewing the decision and its consequences.

Other Internal and External Factors

Mealy (1984) identifies eight types of factors that influence client decision making:

- Medical constraints: information about the disorder, availability of medical options, client health, and failure rates of tests and procedures.

- Financial constraints: limited financial means, uncertainty about future financial status, and insurance coverage.

- Legal and societal values: obvious societal restrictions on available options (e.g., abortion) and subtle restrictions such as funding for research, social programs, and services for dealing with a particular condition.

- Family values: the nature of family members' values and the ways in which they express them can either be a support or a stress.

- Client motivation: factors affecting client desire and ability to make a decision such as intelligence, education level, stress levels, and willingness to participate in genetic counseling.
- Client values: personal attitudes and values about one's options (e.g., contraception, having a normal child, etc.) as well as about taking personal responsibility for the decision, attitudes about medical personnel, concern for others' feelings, etc.
- Client personality: affects how the client approaches decision making, for example, compulsively, fearfully, dependently, etc.
- Counselor constraints: what the counselor is able and willing to provide; and legal considerations, rules and policies, and counselor training and values.

Mealy points out that these eight factors can either hinder or facilitate decision making and that genetic counseling outcomes are influenced by an interaction between client characteristics and behaviors and genetic counselor characteristics and behaviors. Mealy further notes that a factor that helps decision making may hinder subsequent adjustment. Consider, for example, a prenatal couple. The husband makes all of the decisions, and his decision to terminate the pregnancy happens fairly quickly and easily. Subsequently, his wife experiences extreme grief and resentment over his decision.

Cassidy and Bove (1998) identified four themes related to parents' decisions about whether to seek or reject presymptomatic testing for their children who are at risk for adult-onset genetic conditions that are treatable: (1) personal experience with the severity of the genetic condition, (2) receipt of accurate information from credible sources, (3) availability of treatment, and (4) risk perceptions.

Sagi et al (1998) identified a client's knowledge base as one factor that impacts problem solving and decision making. This knowledge base includes information, personal experiences, and beliefs that are at one's disposal.

Another factor is accessible information, which is influenced by both the genetic counselor's and the client's experience. For example, Wroe and Salkovskis (1999) pointed out that the information that you present and the factors that you emphasize affect your client's likelihood of pursuing genetic testing, perceptions of severity of the condition, and level of anxiety. Even counselors who believe that they are taking a nondirective approach can subtly influence client decision making by influencing the factors that are most accessible at the time of the decision. Furthermore, even if it were possible to lay out all of the conceivable factors, there is no way that you can make clients give equal amounts of time or attention to them. They argued that clients will make decisions based upon *accessible* information, which is not necessarily the same as relevant information. An example of accessible information would be a client's knowledge that her neighbor had a healthy baby when she was 40, so the client decides not to pursue prenatal testing; this information does not affect the client's own advanced maternal age risk at 40. Wroe and Salkovskis further argued that individuals make decisions based on subjective premises that may not be shared by others; thus, clients make a decision based on what they believe is relevant at the time: "These premises may be totally idiosyncratic, may actually be factually wrong, and may also be systematically influenced by other factors (such as anxiety)" (p. 21). They also found that individuals who focus on negative aspects of their susceptibility may engage in a cognitive defense of minimizing (e.g., It's not such a bad condition; I probably don't have the gene, etc.). Therefore, one challenge you will face is that some clients will not focus on all of the factors that you regard as pertinent to their decision-making situation.

Reproductive Decision-Making Factors

Van Spijker (1992) identified nine factors that influenced the reproductive decision making for a sample of males and females. In descending order of frequency for both genders, they are:

- recurrence risk
- desire to have a child
- availability of prenatal diagnosis
- coping skills
- impact of the disorder
- family factors (e.g., finances, other resources)
- diagnosis
- norms and values
- reproductive alternatives

She also found that after genetic counseling, 30% of the couples changed their risk perception, 15% were unsure about a next pregnancy, and 14% of the couples disagreed about future reproductive plans.

Frets et al (1992) found that the availability of prenatal diagnosis significantly affects reproductive decisions. When prenatal diagnosis was possible, 87% of their sample decided to have children; when prenatal diagnosis was not possible, only 47% decided to have children. Furthermore, couples who had the option of prenatal diagnosis were likely to regard their decision-making process as more burdensome than did couples for whom prenatal diagnosis was not available. The authors concluded, "The technology of prenatal diagnosis did not provide the easy way out" (p. 23). They also found that reproductive decisions were complicated for couples by (1) difficulties during the decision-making process, (2) doubts about the decision they made, and (3) inability to make a decision. Over half of the couples in their study experienced guilt, especially those couples who had an affected sibling. These findings suggest that "conscious and unconscious affective issues are important in the decision-making process" (p. 25).

Lippman-Hand and Fraser (1979a) identified three aspects of the decision-making process that need to be considered when couples are involved in reproductive decision making:

- Ambiguity: the uncertainty concerning what a particular diagnosis would mean for the health and functioning of an affected child (e.g., how severe would Down syndrome be?).

- Burden: associated with the need to make a decision (e.g., knowing that a future pregnancy could result in another affected child increases the burden about deciding to become pregnant).

- Desire to normalize: wanting to prove their ability to produce healthy offspring.

Obstacles to Decision Making

Decisions always involve both cognitive and emotional factors (Wroe and Salkovskis, 1999). These factors vary in the extent to which they are grounded in reality (rational) or based on clients' mistaken perceptions (irrational). These factors can present obstacles. For example, "We don't want to disappoint others"; "We want to know all of the information for certain"; "We don't want to make a mistake"; "We don't want to be responsible for the outcome."

McCarthy Veach et al (2001) interviewed genetic counselors, physicians, and nurses who reported three challenges that pose difficulties in decision making:

- Lack of informed consent: patients don't know, don't want to know, and/or don't understand all of the pertinent information.

- Facing uncertainty: lack of test specificity and sensitivity, and the reality that no one can know all of the short- and long-term outcomes.

- Disagreements: with family members, cultural groups, health care providers, and society about what to do. There is seldom an obvious choice.

A Rational Decision-Making Model
for Genetic Counseling Clients

Because some clients will feel "blocked" in their decision-making process, it may be helpful if you offer them an opportunity to think through their situation (Kessler, 1997). One option is a decision-making model such as the Danish and D'Augelli (1983) model that is based on three major assumptions: (1) it is useful to break decisions down into relevant factors and to weigh each factor according to each alternative (option); (2) a systematic model helps to decrease anxiety about a decision and helps to put the factors involved in a decision into greater perspective; and (3) a rational model includes emotions as a relevant factor, which helps clients deal with the vagueness that feelings often bring to decision-making situations. Their model includes the following steps (which can be written down as they are discussed with a client):

- Client briefly describes her or his situation and decision(s) to be made.

- Brainstorm or consider all possible alternatives (options).
- Brainstorm all possible relevant factors (e.g., medical, familial, cultural, psychosocial, financial, ethical, values, etc.). Try to be as comprehensive as possible, including short- and long-term factors, and break global factors into specific ones (e.g., feelings could be broken down into relief, guilt, anger, depression, etc.). Try to bring to the surface irrational factors that clients may be reluctant to acknowledge (e.g., "Having an affected child might mean that I'm a failure as a parent").
- Evaluate each alternative. Does it satisfy each factor?
- Client chooses the most important factors.
- Client determines the most desirable and best alternatives. The desirable alternative is the one that attracts the client the most; the best alternative is the one that is in the client's interest in the long run, although it may not be desirable at first glance.
- Clarify and review the decision and revise it as new factors arise.

The following example illustrates the Danish and D'Augelli decision-making model:

Client Situation

The client is seeing you for prenatal genetic counseling. She is 36 years old and pregnant, with a 4-year history of infertility. She had two abortions when she was younger. During this pregnancy she initially decided against an amniocentesis, and instead opted for a triple screen. The results indicate an increased risk for Down syndrome of 1/100. In talking with the client, you determine that the major decision she is trying to make is whether or not to have an amniocentesis. Together, you identify the following factors that are influential in her decision:

Decision

Proceed with amniocentesis (alternative A) or do not have amniocentesis (alternative B). See table page 138.

Evaluate Each Factor

In the above table, the client and counselor identified each factor that would lead to alternative A or B by placing an X in the column of the alternative most likely to satisfy or prevent each factor.

Choose the Most Important Factors

Upon further discussion, the client identified the factors that are most important to her decision: risk for Down syndrome, anxiety over the triple screen, and her religious belief that abortion is wrong.

Relevant factors	A	B
Medical:		
Risk of miscarriage		X
Risk for Down syndrome	X	
Psychosocial:		
First and possibly only pregnancy with husband		X
Guilt stemming from previous terminations		X
Irrational belief that she is being punished for previous terminations		X
Anxiety over triple screen results	X	
Grief over termination—losing the pregnancy		X
Her parents strongly oppose abortion for any reason		X
Husband's strong disapproval about past terminations		X
Relief of knowing for sure	X	
Financial:		
Burden of caring for a child with special needs	X	
Cannot afford additional infertility treatments		X
Meaning of abnormal results:		
Feeling responsible for having a defective child	X	
Facing difficult decision of terminating or maintaining the pregnancy		X
Risking having no children (if can't become pregnant again)		X
Client's religious beliefs:		
Against abortion		X
Viewing this as a test of God's faith, and failing that test by considering abortion		X

Determine the Most Desirable and Best Alternatives

The client stated that alternative B is the most desirable option, but alternative A is the best alternative. To help the client move forward in her decision-making process, you could work with her to minimize some of the factors by suggesting that she talk things over with her husband before making a decision and by pointing out that a decision to have an amniocentesis does not mean that she has made a decision to terminate the pregnancy if the results are abnormal. Other possible options to consider are having a level II ultrasound instead of an amnio, although this does not eliminate the risk, and adopting if results are abnormal. The client can also read some literature about Down syndrome.

When using the Danish and D'Augelli model, you should watch for these potential pitfalls:

• Many clients will limit their thinking about alternatives (e.g., termination vs. have an affected child may be all that they identify). Adoption might be another option, for example.

• While brainstorming factors, clients may jump prematurely to evaluating alternatives for the factors. Try to stop them from doing this as it usually prevents clients from identifying all relevant factors.

- Clients may be reluctant to admit to irrational factors. You may need to tentatively suggest some, such as choosing an option that others want you to choose (having an abortion because your parents want this); worrying about the impact on your self-esteem (having an abortion to keep yourself from feeling responsible for giving your child a birth defect); superstitious beliefs (e.g., if you test for a genetic condition, you will bring on the disease).

- Probably no alternative will satisfy all important factors. If it did, then the decision would be clear. When all alternatives present seriously negative consequences, then you might talk with your client about which option is least risky, as perceived by the client. Another strategy, which comes from Lewin's (1947) force-field analysis approach to decision making, is to talk through how your client could eliminate or reduce the negative consequences for different factors. This may allow an alternative to become more desirable. For example, in the previous scenario the client could go home and talk to her husband to check on whether his negative reactions to her previous terminations would generalize to this situation; and a level II ultrasound might decrease some of her anxiety about the triple screen.

Some Suggestions for Assisting Clients in Their Decision Making

Regardless of the decision-making model that you use with clients, the following strategies can help you assist them in their decision-making process:

- Reassure clients that they have the ability to make the best decision for themselves.

- Convey understanding and acceptance of your clients no matter what decision they make. This is essential to nondirective genetic counseling (Bartels et al, 1997).

- Clients tend to approach genetic counseling decisions the way they typically have approached big decisions in their past. So it may be useful to ask your clients to briefly describe their typical decision-making style (e.g., "Think about an important decision that you have made in the past. How did you go about making the decision?" or, "How does the way you're going about making this decision compare with the way you've made other big decisions in your life?").

- If clients are going about making a decision in an atypical way, for example, rational deciders who suddenly become very dependent and want you to tell them what to do, this can be important evidence that they feel overwhelmed, or that some important factor is in the way and needs to be discussed. Consider pointing out this discrepancy and talking about how they are feeling.

- Clients do not have to make a final decision on the spot. Even in prenatal counseling, where there may be a greater time pressure, usually a client or couple can go home and sleep on it.

- Keep in mind that you have not failed as a genetic counselor if your client doesn't make a decision. Remember, not deciding is a decision. It's a decision not to decide. This strategy lets time and/or circumstances choose the option (e.g., putting off a decision regarding termination of a pregnancy means the time will come when termination will not be an option, except in extreme circumstances). However, it is important that you point out these consequences to indecisive clients.

- Explore with clients their reasons for making the decision. For example, "What will choosing this option mean for you?"; "What do you see as the pros? The cons?"; "Sometimes clients choose this option because it will... Is it possible that this is part of what is motivating you to make this choice?" Clients need to be honest about their motivation, and they must determine whether this is the motivation they wish to have driving their decision (e.g., a client who is rushing into a decision before all of the pertinent information is available because she does not like how anxious she feels).

- You will need to explore differences in opinions and attitudes when your client is a couple or family (Van Spijker, 1992). Go around the group and ask each individual to express what s/he thinks and feels about each option.

- Recognize and incorporate cultural variables in the decision-making process. For instance, for some Korean and Arab families, it is important to involve the fathers as they are the decision makers (Brown, 1997).

- Work to understand both conscious and unconscious guilt feelings that clients may have toward affected family members (Frets et al, 1992). You will need to watch for client nonverbal behaviors that suggest guilt and listen to what is said and not said (see Chapter 8).

- Let clients know that sometimes the most emotionally painful decisions are the right ones for them. The best decision is not necessarily the easiest to make.

- Suggest that clients listen to their instincts. Often our subconscious is a source of good advice. This may be particularly helpful for clients who are intellectualizing their situation, that is, spending all of their time thinking things through without acknowledging how they feel about different options.

- Encourage clients to seek support and guidance from significant others (e.g., family members, friends, community leaders).

- Engage in anticipatory activities:

  — Huys et al (1992) and Van Spijker (1992) suggest the use of scenarios. This can be particularly helpful when clients are unfamiliar with a disorder and/or have no family history with the condition. For reproductive decisions, after presenting the scenario, Van Spijker suggests that you ask three questions [from the study by Lippman-Hand and Fraser (1979b)]: (1) How likely am I to have an affected child? (2) What will it be like if it happens? (3) How will others react to my choice?

  — Frets et al (1992) describe scenarios as constructing a plausible story, in which the decision maker is an active participant. Scenarios describe what could happen or could be done under various conditions (e.g., having an affected child vs. having a healthy child). They point out that you will gain valuable clinical information about how clients represent and reason through information based on the scenarios that they construct. In their research, Huys et al (1992) found that people typically construct between three and eight scenarios, and that the contents of scenarios are quite divergent, indicating that they are highly personalized. These findings suggest that your clients won't try to come up with every possible outcome, but instead will focus on a few outcomes that are particularly important to them.

— Bottorff et al (1998) recommend using predisclosure role-plays. These are exercises that invite clients to consider the effect of test results on themselves and their family members. For example, you could ask, "If the results are positive, what do you think this will mean for you? How do you think you will feel? What will you do?" Then ask your client to answer the same questions, but for specific family members (spouse, children, etc.). Finally, ask these same questions, but with your client imagining that the results are negative.

— Kessler (1997) suggests having clients role-play or pretend that they are coping with a specific situation or person. This allows them to try out different strategies/alternatives/options. For example, role reversal might be helpful for a couple who is disagreeing about a reproductive decision, allowing them to see things from each other's perspective.

• Consider referring undecided clients to psychologists or to others who are familiar with the specific difficulties they are having in the decision-making process (Frets et al, 1992).

• Provide a secondary support, that is, provide additional or repeated information, and make a referral to psychological counseling or other sources to assist clients in their adjustment to the outcomes of their decisions (Mealy, 1984).

• Ask yourself, what is a good decision? McCarthy Veach et al (2001) found that most of the genetic counselors they interviewed felt comfortable when their clients made decisions that would not cause them harm, were consistent with their cultural backgrounds, and seemed to work for those particular clients. The hardest decisions were those perceived by the genetic counselors to be cavalier (e.g., terminating a twin pregnancy after years of infertility because the couple wanted only one child). One major ethical/professional challenge that you will face is refraining from talking clients out of decisions with which you disagree. (In Chapter 11 we discuss nondirectiveness in greater detail.)

Closing Comments

In this chapter we suggested several strategies to help you walk with your clients through the difficult genetic counseling process. First, you will provide relevant information to help them be informed. Then, for many clients, you will help them use this information and other relevant information to come to a decision that is best for them. Decisions are complex. Clients select certain options for numerous reasons, some clear, some not so clear. And they often choose options that are different from what you think they should choose. Remember that what seems to be an irrational decision to you can be the best choice for your client. It is also important to remember that you can't fix situations that aren't repairable—some clients' decisions come down to trying to do the best they can in an impossible situation.

Class Activities

Activity 1: Dyads and/or Small Group

Students brainstorm words that could communicate a negative message or stigmatize the client when presenting risk information (e.g., mutant).

PROCESS

The whole group generates a list that the instructor writes on the board.

Estimated time: 15 minutes.

Activity 2: Dyads and/or Small Group

Students discuss how they will feel when communicating abnormal test results to clients. What is difficult about it? What is scary about it? What is the worst thing that could possibly happen?

PROCESS

The whole group discusses their responses.

Estimated time: 20 to 25 minutes.

ACTIVITY 3: DYADS AND/OR SMALL GROUP

Using the risk data for the following genetic conditions, students write out how they would communicate these data to a client using Hallowell et al's (1997) five quantitative formats:

- Choroid plexus cysts: controversial association with trisomy 18 and 21. Risk is probably less than 1%.
- Sickle cell anemia: father of the baby is a carrier. Carrier risk is 1/10 for African-American population. Explain risk for being a carrier, and then explain risk for having an affected child.
- Neural tube defects: 3% to 5% empiric recurrence risk.
- Abnormal alpha-fetoprotein (AFP) with increased risk for Down syndrome: 1 in 50.

- Newborn diagnosed with abnormal hemoglobinopathy. Follow-up needed.
- Hypoplastic left heart. General population risk: 16/1000. Risk for recurrence if one affected child: 2% to 25%.
- Cystic fibrosis (CF) carrier risk (Caucasian): 1/25. Carrier risk if screening negative: 1/265.

Estimated time: 45 minutes.

Instructor Note

- After writing down their responses, students could role-play actually giving the information to clients.

Activity 4: Dyads and/or Small Group

Students brainstorm the reasons that it is hard to make decisions.

Process

The whole group discusses the reasons that they generated.

Estimated time: 15 minutes.

Activity 5: Dyads and/or Small Group

Students think about a major decision they have made in their lives (e.g., choice of graduate program). How did they go about making this decision? Next the students try to match their process to one of Dinklage's (1966) eight decision-making styles. Then they discuss the advantages and drawbacks to using their own decision-making styles when working with clients.

Estimated time: 20 minutes.

Activity 6: Small Group Role-Plays

Each student takes a turn playing a genetic counselor and client. The client selects one of Dinklage's (1966) eight decider types and demonstrates that style for a decision about whether or not to have an amniocentesis. The client should not let the group know in advance which style s/he chooses. Each role-play should last for about 10 to 15 minutes.

Process

The students discuss how it feels to deal with different types of deciders.

Estimated time: 90 minutes.

Activity 7: Small Group

Using one of the following client scenarios, first go through the Danish and D'Augelli (1983) decision-making steps and then discuss how to minimize constraining factors.

The client is coming for genetic counseling because of a family history of mental retardation. Her maternal uncle and some other maternally related males are mentally retarded. She is terrified of having a baby who is mentally retarded and is adamant that she does not want to be in that situation. She definitely wants an amniocentesis and wants every possible test done on the fluid. She is not concerned about the cost of any of the tests.

OR

A 15-year old African American is 2 1/2 weeks into her first pregnancy (g1P0) and has just been told that her fetus has gastroschisis. No other anomalies or abnormalities were found on the sonogram. As her mother is a severe drug addict, her grandmother has accompanied her to the clinic. The grandmother is adamant that the pregnancy must be terminated.

OR

A 32-year old woman has two sons; both have Fragile X syndrome (one mild, one more severe). She wants another child. Her husband is supportive and involved, but doesn't want any more children.

PROCESS

If there is more than one small group, each group can present its decision model to the other groups. Then the whole group discusses any questions/difficulties they are having with the model.

Estimated time: 75 to 90 minutes.

Activity 8: Triads and/or Small Group

Role-play presenting risk information to:

- A 24-year-old prenatal client with neurofibromatosis who has limited cognitive skills.
- A 39-year-old prenatal client with a family history of CF who says, "I don't want to hear anything about risks because it scares me. Just tell me what to do."
- A Muslim refugee from Africa in a consanguineous relationship who states that, "How children turn out is God's will, and there is no way to predict or prevent birth defects."

Estimated time: 45 to 60 minutes.

Activity 9: Modeling

The instructor role-plays for 45 minutes with a volunteer from class. The instructor demonstrates how to (1) present risk information; (2) present abnormal test results; (3) assist a client with decision making.

After the role-play students discuss their observations of the counselor's behaviors and their impact on the client.

Estimated time: 90 minutes.

Activity 10: Small Group Role-Play

The students form small groups (four to five students). In each small group, either the instructor or a student pretends to be a genetic counseling client. The client role-plays one of the following scenarios that involve client decisions:

- The prenatal client states that she does not want to consider abortion as an option before the genetic counselor has said anything about it.
- The client wants the genetic counselor to tell him what to do regarding whether or not to have presymptomatic testing for colon cancer (familial adenomatous polyposis, FAP).
- The client minimizes a "large" risk factor for cancer when she carries the BRCA-1 gene.
- The client misunderstands the risk information given to her (she thinks having a 25% chance of passing a gene for cystic fibrosis to her children means that only one of four of her children would have the gene).
- The client with breast cancer greatly overestimates her risk for having a gene mutation (90% when her risk is actually 5% to 10%).
- The client is afraid of what her family will think if she decides to terminate the pregnancy.

The role-plays can last 15 to 20 minutes. The counselor or instructor can stop a role-play if the counselor gets stuck, and discuss with the group how to proceed.

Estimated time: 2 hours.

Activity 11: Triad Exercise

In 45-minute role-plays, three students practice (1) giving risk information, (2) giving abnormal test results, and (3) decision making. They take turns as counselor, client, and observer. They spend 10 minutes for feedback after each role-play. The observer should stop the counselor if s/he is stuck.

In the large group, students discuss what they learned from the role-plays, what questions they still have about information giving abnormal test results and assisting in decision making.

Estimated time: 3 to 3 1/2 hours.

Written Exercises

Exercise 1

The following scenarios provide information about particular clients and their genetic risks. For each scenario write down different ways that you could communicate risk to the client.

1. The client is a 43-year-old woman, 11 weeks pregnant. This is her first pregnancy; she has a history of infertility for 7 to 8 years. She works as an elementary school teacher. Her husband is a foreman for a plumbing company. He is of Italian descent; she is of Irish/English descent. She was referred by her obstetrician.

2. The client has a familial history of breast cancer. Her mother and maternal aunt died in their 50s. The client, in her early 30s, wants to know her risks.

3. A young couple comes for prenatal counseling. The man's sister has a son with Lesch-Nyhan syndrome. He wants to know if his children are at risk of having the disorder.

4. The same young man in scenario 3 sends his sister (the one with the affected son) for counseling. She has just moved to the United States and has not previously had access to genetic counseling. She wants to know if her other children will develop Lesch-Nyhan, and also what are the risks to future pregnancies.

5. The client is coming for genetic counseling regarding a history of two babies who died shortly after birth. A genetics consult was ordered on the second baby, and the client is here to review results and recurrence risks with you. No specific diagnosis was made; however, the recurrence risk is most likely 25%, and the client is stunned and very upset that the risk is so high.

6. The client is a 25-year-old woman whose male child died in the newborn period from complications of multiple congenital anomalies. Based on exam and autopsy, it is most likely that the child had a genetic syndrome, but the child's features do not fit any previously described disorder. In addition, the karyotype was within normal limits. The client is no longer with the birth father. She is interested in discussing her risks for having a similarly affected child.

INSTRUCTOR NOTE

- One variation is to have students select one or more of the client scenarios and write them as if they were presenting the information to two or more of the following clients:

A statistician

A judge

A physician

A special education teacher

A mentally retarded client

An adolescent

Or, you can create any type of client for whom you wish them to prepare the risk explanation.

- These scenarios could be role-played in class, with the student verbally presenting risk information to the client.

Exercise 2

For one or more of the scenarios described in exercise 1, identify the client's possible options, and all factors that might be relevant to the client's decision, using the Danish and D'Augelli (1983) decision-making model.

Exercise 3

Observe an actual genetic counseling session and afterward use the Danish and D'Augelli decision-making model to describe the client's decision-making process.

Exercise 4

Talk with an individual who is a recognized cultural leader for a specific population (e.g., Hmong, Native American, Latino/Latina, etc.). Discuss how members of her or his community typically understand the concept of risk and how risk can best be communicated.

Exercises 5

Engage in a 60-minute role-play of a genetic counseling session with a volunteer from outside the class. During the role-play focus on providing risk information, presenting abnormal test results, and assisting the client with decision making. Audiotape the role-play. Transcribe the role-play and critique your work. Give the tape, transcript, and self-critique to the instructor who will provide feedback. Use the following method for transcribing the session:

Counselor	Client	Self-Critique	Instructor
Key phrases of dialogue	Key phrases	Comment on your own response	Provides feedback on your responses

Exercises 6

Simonoff (1998) lists several types of questions that clients may have when they seek genetic counseling for autism. These questions, which are generally relevant for prenatal genetic counseling, include:

- What are the recurrence risks?
- Is there increased risk for other conditions?
- Was the diagnosis accurate?
- Would other family members have the same type and level of impairment as the proband?
- What types of prenatal diagnosis are available?
- Are any precautions during pregnancy necessary?
- Is the condition more common in males or females?
- What is the impact of another affected child on the proband?
- What is the impact of a nonaffected child on the proband?
- How early can the condition be detected?
- Would risks differ if either partner were to have children with a different person?

Assign each student a genetic condition and have them research the condition, using Simonoff's list of questions for prepping the case.

INSTRUCTOR NOTE

- Students could then role-play these cases in class.

Annotated Bibliography

Arnold, J. R., Winsor, E. J. T. (1984). The use of structured scenarios in genetic counseling. Clinical Genetics, 25, 485-490.

[Describes a method for assisting clients in making decisions based on their genetic risks.]

Bottorff, J. L., Ratner, P. A., Johnson, J. L., Lovato, C. Y., Joab, S. A. (1998). Communicating cancer risk information: the challenge of uncertainty. Patient Education and Counseling, 33, 67-81.

[Describes the complex dimensions of communicating risk information.]

Cassidy, D. A., Bove, C. M. (1998). Factors perceived to influence parental decision-making regarding presymptomatic testing of children at risk for treatable adult-onset conditions. Issues in Comprehensive Pediatric Nursing, 21, 19-34.

[Discusses the factors important to parental decision making about whether to seek or reject presymptomatic testing for their children.]

Danish, S. J. D'Augelli, A. R. (1983). Helping skills II: life development intervention. New York: Human Sciences Press.

[Describes a decision-making model in detail and includes examples and practice exercises.]

Fisher, N. L. (1996). Cultural and ethnic diversity: a guide for genetics professionals. Baltimore: Johns Hopkins University Press.

[Describes possible cultural differences that impact information-giving and decision-making processes.]

Jecker, N. S., Carrese, J. A., Pearlman, R. A. (1995). Caring for patients in cross-cultural settings. Hastings Center Report, 25, 6-14.
[Discusses cross-cultural differences between professionals and patients that affect the quality of care provided.]
Kessler, S. (1987). Psychological aspects of genetic counseling. IV. The subjective expressions of probability. New England Journal of Medicine, 315, 741-744.
[Discusses the variability across genetic counseling clients in how they interpret/perceive numerical risk data.]
Mealy, L. (1984). Decision making and adjustment in genetic counseling. Health and Social Work, 9, 124-133.
[Provides a good description of eight factors that can hinder or help genetic counseling client decision making, the relationship between decisions and subsequent adjustment, and how genetic counselors can facilitate decision making and adjustment.]
Simonoff, E. (1998). Genetic counseling in autism and pervasive developmental disorders. Journal of Autism and Developmental Disorders, 28, 447-456.
[Provides several examples of how to present risk information and includes a table of questions that clients might ask their genetic counselors.]
Wroe, A. L., Salkovskis, P. M. (1999). Factors influencing anticipated decisions about genetic testing: experimental studies. British Journal of Health Psychology, 4, 19-40.
[Discusses how the way in which genetic counselors present information either positively or negatively influences client likelihood of pursuing testing, perceptions of severity, and anxiety levels.]

CHAPTER 8

Responding to Client Cues:
Advanced Empathy and Confrontation

Learning Objectives

1. Define advanced empathy and confrontation.
2. Differentiate advanced empathy and confrontation from primary empathy.
3. Determine guidelines for effective advanced empathy and confrontation.
4. Identify examples of client themes appropriate for advanced empathy and confrontation.
5. Develop advanced empathy and confrontation skills through practice and feedback.

This chapter discusses two basic helping skills: advanced empathy and confrontation. Typically, genetic counselors use these two types of responses less frequently than other responses such as primary empathy and questioning. Advanced empathy and confrontation can be very powerful responses when used strategically and sparingly.

Advanced Empathy Skills

Definition and Functions of Advanced Empathy

Advanced empathy is a helping skill that consists of two components: (1) the genetic counselor's understanding of the underlying, implicit aspects of client experience; and (2) the response or reply the counselor constructs based on this understanding. Advanced empathy responses go beyond surface client expressions by identifying less conscious client feelings, thoughts, and perceptions. Advanced empathy is a tentative hypothesis or guess about the client's experience based on evidence from the client (Turock, 1980).

Clients often come to genetic counseling with a vague awareness of their inner thoughts and feelings. Even if they have a pretty good idea of what they think and feel, they may be hesitant to share this information because of fears of being judged or beliefs that what they have to say is too risky (Turock, 1980). When you have reason to believe that there is more beneath the surface of your clients' stories, advanced empathy can be helpful because it can offer more direct interpretations of their inner experience.

With advanced empathy responses, you go beyond what the client has directly expressed by presenting your perspective of her or his experience. You move from client descriptions of their experience to a more objective stance to offer a new meaning or reason for their feelings, thoughts, or behaviors (Hill and O'Brien, 1999; Kessler, 1997). Your intent with advanced empathy is "not to deny the way the client sees the world, but to present the client with an expanded view of the world" (Geldard, 1989, p. 72). With advanced empathy you become more directive about the discussion, having made a decision that your client would benefit from hearing your perspective. Because advanced empathy addresses hidden or implied client content, it can increase client anxiety (e.g., "Will the counselor judge me now that s/he knows this about me?"; "Do I want to get into this with the counselor?"; "Will I be able to handle this?"; etc.). Advanced empathy tends to occur later in the genetic counseling session once you have developed trust and rapport with your client.

Chapter 4 presented primary empathy on a continuum ranging from silence to reflections of content and affect. If we extend that continuum, advanced empathy would be to the far right:

The Primary and Advanced Empathy Continuum

Silence	Minimal Encourager	Paraphrase	Summary	Reflect Content	Reflect Affect	Content and Affect Reflection	Advanced Empathy

Primary and advanced empathy differ in a number of ways:

Primary Empathy	Versus	Advanced Empathy
• Interchangeable or synonymous with client's explicit expressions		Additive—goes beyond client's explicit expressions
• Deals with surface content and affect		Deals with hidden, implied content and affect
• Reflects client point of view		Reflects counselor point of view
• Counselor is responsive to discussion		Counselor takes initiative to direct discussion
• Client is more aware of the feelings and thoughts reflected by the counselor		Client is less aware of the the feelings and thoughts until counselor reflects them
• Reassures client		Challenges client
• Lowers client anxiety		Raises client anxiety
• Clarifies and builds trust		Provides insight and promotes change
• May occur throughout session		Usually occurs later in the session
• Used frequently		Used sparingly

Advanced empathy can provide clients with greater insight into their thoughts, feelings, and values and lead to greater self-understanding. It can also give clients permission to express certain feelings or opinions, which may ultimately help them move toward self-acceptance and facilitate their goal setting and decision making. Because advanced empathy can provide clients "with a conceptual framework that explains their problems and offers them a rationale for overcoming their concerns" (Hill and O'Brien, 1999, p. 207), it challenges the client to develop greater insight and possibly to change behavior. Advanced empathy is sometimes referred to as interpretation, reframing (Kessler, 1997), and additive empathy. We prefer *advanced empathy*, and will use this term almost exclusively throughout the chapter.

Psychotherapy research consistently has demonstrated that advanced empathy has a positive impact on process and outcome; for example, it enables clients to experience their feelings, and they report favorable reactions to its use (Hill and O'Brien, 1999). Researchers have indicated that advanced empathy occurs less frequently than primary empathy, comprising only about 6% to 8% of psychotherapist responses (Hill and O'Brien, 1999). Since advanced empathy is such a powerful response, you should use it sparingly.

Guidelines for Using Advanced Empathy

Effective advanced empathy requires accurate understanding and sensitive responding to clients. We recommend the following strategies for formulating and communicating advanced empathy responses:

GENERATE HYPOTHESES ABOUT CLIENT SITUATIONS, THOUGHTS, AND FEELINGS

- Review the file if you have access to client information prior to the genetic counseling session. Spend a few minutes formulating tentative hypotheses based on client demographics (age, gender, etc.), medical data, and reasons for seeking genetic counseling.

- Look for cues when you first meet a client (e.g., How relaxed or tense is the client? Whom did the client bring along to the session?).

- Use past experiences with genetic counseling clients and your knowledge of psychosocial theories to anticipate underlying client affect and content.

- Use your own experiences. Put yourself in the client's place and ask yourself how you might feel if you were this client. But be careful not to project your feelings onto the client.

- Pay attention to client verbal and nonverbal behaviors.

- Listen for themes and repetitive patterns. Novices often make the mistake of thinking that different pieces of information only go together if the client talks about them at the same time (Mayfield et al, 1999). In reality, clients may provide related information at different points in the session, so you will have to fit the pieces together to see the themes.

- Ask yourself, "What is my client trying to tell me that s/he can't say directly?" (Turock, 1980).

- Remember that cultural and individual differences mean that no two clients will understand the same experience exactly. Don't go overboard with theories that fail to match your client's experience. Identifying feelings or thoughts incorrectly can be worse than saying nothing. Additionally, it is important to listen for the client's understanding of illness within her/his cultural context: "One must identify the patient's interpretation of the physician's [or genetic counselor's] explanation and also explore their non-medical beliefs about the disease so as to be helpful to the patient" (Handelman et al, 1989, p.471).

- Cultural differences influence clients' "health beliefs and practices, social customs, family structure, attitudes toward medical personnel" (Weil and Mittman, 1993, p. 160). Furthermore, some non-Western clients, who seemingly have acculturated to Western customs, will revert to old beliefs and practices when they are in crisis (Weil and Mittman, 1993).

SHARE YOUR HYPOTHESES THROUGH CAREFULLY FORMULATED RESPONSES

- Be concise, clear, and specific.

- Use responses that are nonjudgmental and nonpresumptive.

- Be tentative. Allow a client the chance to deny or modify your statement. Frame advanced empathy "with an attitude of curiosity, of wondering what it is that makes the client act a certain way" (Hill and O'Brien, 1999, p. 211). For example, "Correct me if I'm wrong, but it seems that you're saying..." You can also lead up to advanced empathy by first asking for the client's interpretation (Hill and O'Brien, 1999).

- Formulate responses that are moderate in depth. Several psychotherapy studies indicate that interpretations that are of a moderate depth rather than too superficial or too deep have the most positive effect on process and outcome (Hill and O'Brien, 1999). Refrain from jumping in with dramatic interpretations that will be off-putting to your clients.

- Be sure that your response is suitable for a given client. One way in which clients differ is in their degree of psychological-mindedness. Some clients are more psychologically minded than others and respond well to interpretations about their inner experience; other clients are less interested in the why of their experience, and more interested in support, in answers, and/or in feeling better (Hill and O'Brien, 1999). Clearly, you would use fewer advanced empathy responses with the latter type of clients: "Because our approach is essentially client-centered and we value empathy above all, it is important not to impose our value for insight on such clients but to respect their choice not to understand themselves" (Hill and O'Brien, 1999, p. 185). This perspective from psychotherapy seems appropriate to genetic counseling as well. Another way in which clients differ is in how trusting they are. Some clients are extremely mistrustful and suspicious; you have to stay close to the surface with them, using primary empathy (Martin, 2000).

- Give a well-timed response. Usually you will make an advanced empathy statement only after you've built some rapport, for example, with primary empathy (Josephs, 1997) and when you have enough impressions about the client to be able to trust your hypothesis. Make advanced empathy statements when clients seem to be ready (i.e., have clearly stated their concerns, stated that there are some things they do not understand and seem eager to understand, and when they have a high level of emotional distress that is prompting them to find a solution to their concerns) (Hill and O'Brien, 1999). Also, you should anticipate how your client will react to your interpretation before giving it (Martin, 2000).

- Use advanced empathy infrequently. Be sparing in your use of this powerful response. Clients can usually deal with a limited number of insights at one time because insights may have a strong emotional impact. Psychotherapy research (e.g., Olsen and Claiborn, 1990) has demonstrated that interpretations elicit greater client anxiety than do primary empathy statements.

- Observe the extent to which your advanced empathy was accepted by your client. Your advanced empathy has likely missed the mark if your client rejects what you said, becomes silent and withdrawn, or quickly changes the subject (Martin, 2000). Watch your client's reactions to assess whether your advanced empathy was accurate, well timed, and not too threatening. Possible client reactions to advanced empathy include (1) agreeing with your interpretation and exploring its meaning; (2) agreeing, but avoiding any further exploration; (3) asking for further information about your basis for making the statement; and (4) denying the accuracy of your statement (Turock, 1980).

- Follow up with a primary empathy statement. Reflect your client's reaction to and thoughts about your advanced empathy statement.

Types of Advanced Empathy Responses

There are several types of advanced empathy responses that you might use:
- Reflections of feelings and content not directly stated by the client. Example: Observe client nonverbal behaviors (clenched fists, red face) and comment, "I noticed that your fists are clenched pretty tightly...like you might be angry."

- Reflections of feelings that underlie emotions that the client has expressed. Example: "You say you are angry, but I wonder if you are also hurt."

- Clear and direct statements about experiences the client is guarded or confused about. Example: "You keep saying that if the result is positive, you'll have to do something about it. Do you mean terminate the pregnancy?"

- Statements that summarize earlier feelings and content into a meaningful whole. Example: "You've said that since your diagnosis you've lost your appetite, cry a lot, have trouble concentrating, and feel down a lot. It sounds like you are feeling depressed."

- Descriptions of patterns or recurring themes. In describing genetic counseling sessions, Baker (1998) recommends, "If a client repeatedly brings up a concern, an issue, or a question, this area should be given special attention" (p. 67). Example: "You keep mentioning how your family will react when they find out about your condition. Are you afraid that they will not want to have anything to do with you?"

- Connections between various parts of the client's problems. Example: "Perhaps part of the difficulty you're having in deciding about testing for this pregnancy is that you had a miscarriage. You've concerned about the risk of losing another baby."

- Logical conclusions to what the client is saying. Example: "And if you put off having the amnio, then you will be past the point where you will be able to make a decision to terminate the pregnancy."

- Alternative ways for the client to view her or his experience. Example: "You said that finding out you have the gene would be awful. You've also said that it's *pure hell* to be always wondering. Is it possible the test might relieve some of that distress?"

Possible Patterns or Themes to Address with Advanced Empathy

With experience you will begin to recognize themes or patterns that are fairly common to your clients. These generally fall into four broad client categories of nonverbal behaviors, affect, attitudes or beliefs, and defenses:

CLIENT NONVERBAL BEHAVIORS PATTERNS

- Client laughter when discussing painful situations: Genetic counseling clients may engage in joking and other forms of levity when they in fact are experiencing intense emotions such as grief, anxiety, or fear. Their laughter may create a safe distance between them and you, it may prevent them from falling apart, or it may hide what they regard as unacceptable emotions. You might say, "I notice that you are smiling, perhaps because you are afraid that you might fall apart right now?"

- Omissions: Listen for omissions of significant information. For instance, a prenatal client does not mention her husband's thoughts and feelings. You say, "I notice that you haven't said anything about your husband's view."

- Other nonverbal behaviors: Watch for nonverbals that indicate there is more beneath the client's calm verbal presentation (e.g., sweating, teary-eyed, trembling chin or hands). Counselor: "You say you're OK, but you look like you're ready to cry."

- Client word choice: Certain words or phrases reveal the feelings and relationships among people (Fine and Glasser, 1996). In genetic counseling, for example, does your client refer to her fetus as a "fetus," "my baby," or "it"? These words can give you clues about the extent to which your client is distancing from or bonding with the pregnancy. Do couples refer to each other by first name or as "the wife" or "him"? These words can provide clues about their level of closeness or hostility.

CLIENT AFFECTIVE THEMES

- Anger: Anger is frequently the surface expression of sadness and grief. Some clients (especially males and clients from some cultural backgrounds) regard certain emotions as evidence of weakness. Anger can be a defense against their perceived weaknesses. You might address their unspoken emotion by saying, "This must be really devastating for you."

- Depression: The feelings underlying depression may be anger, sadness, and despair/hopelessness. Depression typically is a reaction to a real or perceived loss of control. You could address underlying feelings by saying, for example, "It must be frustrating to feel like there's nothing you can do."

- Shame/guilt: Clients who pass on genetic conditions often feel guilt and shame, and clients who have a genetic condition may feel shame about being "defective" or "damaged goods." Counselor: "It seems like you feel that it is your fault that your son has Marfan syndrome."

- Apprehension/anxiety: Most individuals experience at least some anxiety in new situations (e.g., genetic counseling), as well as anxiety about what they may learn. Often, they will not tell you that this is how they feel. Counselor: "I wonder if you feel nervous about being here?"

- Despair/fear: The client feels that there is no solution, no hope, no way of coping. An example of addressing this feeling is, "Are you afraid you won't be able to deal with the diagnosis?"

CLIENT ATTITUDE OR BELIEF PATTERNS

- Clients who view you as the expert and ask you what to do: Although at times it may be appropriate to answer this type of client question directly, most authors (e.g., Fine and Glasser, 1996) caution against becoming an authority for the client to rely on, and the basic tenets of genetic counseling advocate nondirectiveness (Bartels et al, 1997). In many situations you might say, "This decision is so hard that you wish someone would make it for you"; or "It's very hard for you to figure out what to do right now."

- Externalizing beliefs: Some clients may blame others for their situation. For example, "This wouldn't be so hard if I didn't have to wait this long for an appointment with you!" or "I'd be able to decide about having Huntington's testing if my mother didn't get so hysterical every time I mentioned it." We recommend that you sidestep these externalizations as they are very difficult to modify, and instead steer the conversation toward the client, "It sounds as if you've been feeling very troubled about your condition"; or "Do you feel guilty about burdening your mother with your condition?"

- Client believes that fate or destiny or a higher power brought about this situation: Such clients may believe that they are being punished for some transgression (which they usually cannot articulate) (Fine and Glasser, 1996). Furthermore, some cultural groups believe strongly in fate or karma. It is important to assess the extent to which this belief underlies the client's experience. You might say, "I get the impression that you think that having a child with spina bifida is some sort of punishment"; or "I wonder if in your culture, your albinism is considered to be part of your destiny." Later in this chapter we offer additional suggestions for working within these types of cultural perspectives.

- Unrealistic expectations: Some clients believe that they should be able to decide easily and without any distress, or they may think it is silly or abnormal to feel so upset. You could point out the unreasonableness of their expectations. For example, "Maybe you're being a little hard on yourself by expecting to have everything figured out already."

- Feeling too responsible: Clients may blame themselves for every aspect of their situation. For instance: Client: "I would never have miscarried if I'd quit drinking coffee." Counselor: "It almost seems like you're looking for a reason to blame yourself. Are you feeling responsible?"

- Couples or families may want you to take sides: To be effective, you need to remain as supportive as you can toward each participant. You might say, "It seems like you want me to agree with you, perhaps to give more force to your opinion. It's important that you all have a chance to speak and to hear each other." On a related point, do not let clients speak for each other. Indicate at the beginning of the session that it is important to hear from each individual and then during the session invite each participant to speak (Fine and Glasser, 1996). An exception is clients whose cultural practices require that one person do most of the talking.

- Believing their feelings are wrong: You should validate what clients are feeling when their emotions are appropriate to the situation. For example, "It sounds like you have good reasons for feeling frustrated" (Fine and Glasser, 1996).

CLIENT DEFENSES PATTERNS

- Clients who sound as if they are working from a script: Some clients present with "rehearsed stories" (Fine and Glasser, 1996). This may happen if your client has had to repeat the same information to numerous health care professionals, family members, and friends. Try breaking into the script (Fine and Glasser, 1996). For example, you could say, "You must have felt so angry when your father-in-law implied that you shouldn't have any more children." This redirects the client to feelings, away from the rehearsed script.

- Rationalization: The client is trying to justify her or his feelings, beliefs, or choices. You might say, "You keep saying that an abortion would be better for the other children you already have. Based on what you've said about how an affected pregnancy would impact you, I wonder if the abortion would be better for you as well."

- Projection: Clients may attribute their feelings or attitudes to others. For example: Client: "Everyone will think I'm selfish if I terminate this pregnancy because the baby has Down syndrome." In fact, it is the client who feels that she is being selfish. Counselor: "Perhaps you're afraid that you are being selfish."

- Either-or thinking: For example, in a prenatal genetic counseling session, both partners are carriers for cystic fibrosis (CF). They see two options— risk having an affected child or do not have children—because abortion is not an option for them. The couple has not mentioned any other reproductive options. You could introduce the possibility of other options by saying, "So you see only two options, risk having an affected child, or have no children. I wonder if there are any other options you haven't considered."

Challenges in Using Advanced Empathy

Beginning genetic counselors usually find that advanced empathy is a complex and difficult skill to learn to use effectively. Common advanced empathy mistakes include:

- Going overboard with too many interpretations that overwhelm the client. For example, some counselors may need to come across as all knowing, or insightful (Hill and O'Brien, 1999).

- Making advanced empathy statements before clients are ready for them and making your statements too long (Hill and O'Brien, 1999).

- Inaccurately projecting your own experiences onto your clients.

- Lacking "theoretical and personal schema sufficient to provide clients with alternative conceptualizations and hypotheses" (Wilbur and Wilbur, 1980, p. 139). In other words, you lack the experience to be able to see the bigger picture (Mayfield et al., 1999).

- Avoiding using any advanced empathy responses because you fear being wrong about the client; are scared of how the client will react (Turock, 1980); are concerned that you might damage the genetic counseling relationship; or don't want to hurt or embarrass your clients (Hill and O'Brien, 1999). "At some level clients are well aware of their own feelings and perceptions of what has happened to them. We do not need to protect them against the pain of their lives. They have their own defenses to deal with that. More often, they need a witness to hear their pain, their concern, their anger—not someone to change or deflect it" (Fontaine and Hammond, 1994, p. 223).

Some Cultural Considerations in Using Advanced Empathy

In some cultures it is important to be less direct in making advanced empathy responses (Hill and O'Brien, 1999; Pedersen and Ivey, 1993). A less direct approach helps the client "save face." (This approach can be effective with defensive clients as well.)

For example, consider the following subtle ways of addressing client inner experience:

- "In the past when I've had clients in a situation similar to yours, some of them have felt..."
- "Some people might feel [think, do]...if they were in your situation."
- "Some people find it very difficult to...and they choose to..."
- "You say that you're fine with this news, but I want you to know that it's OK if you're not. I hope you'll talk it over with me or someone close to you if you're not comfortable discussing it here."
- "If I were in your situation, I might be thinking about the following...What do you think?"

Generally speaking, it is not a good idea to challenge a person's cultural perspective (e.g., that a genetic condition is God's will). First of all, it is very ethnocentric to believe that your way of viewing reality is better for clients than their own way. Second, clients are quite unlikely to change their perspective on the basis of one or two genetic counseling sessions. Third, this sort of challenge will probably damage any trust that you have established. Try to work with clients within their cultural perspectives. For example, "I understand that you regard your child's metabolic condition as God's will. You may be wondering what I can do for you. If you are comfortable, we can talk about how you might learn whether another baby will have the same thing."

Confrontation Skills

Definition and Functions of Confrontation

Confrontation involves responses in which you directly challenge clients to view themselves and their situations differently. It usually involves behaviors that the client has neither publicly nor privately recognized or acknowledged. Confrontations are a type of feedback that is discrepant with or contrary to the client's self-understanding. Confrontation responses can include identification of client self-defeating behaviors as well as client strengths. Indeed, Kessler (1997) stresses the importance of genetic counselors' identifying "key areas of client functioning which they use throughout the session to strengthen the latter's sense of competence. This might involve parenting, work, interpersonal, or other issues and requires the professional to say rewarding things to the client" (p. 381).

Confrontations are intended to help clients understand their behaviors (thoughts, feelings, or overt actions) and, as a consequence, to consider changing their behavior. Confrontations can challenge discrepancies, contradictions, defenses, or irrational beliefs (Hill and O'Brien, 1999) as well as challenge clients to recognize and use their strengths or potentials. By helping clients explore feelings, attitudes, and beliefs that they have kept hidden, confrontation can remove some of the barriers to goal setting and decision making. In addition, "If done properly, the challenge sets up a mini-crisis that can motivate clients to change. For example, if a helper challenges a client by saying that she looks sad even though she says she is happy, the client is forced to examine the discrepancy" (Hill and O'Brien, 1999, p. 191).

Confrontation shares similarities with advanced empathy since both are counselor-initiated responses that attempt to elicit greater client self-understanding. However, an important distinction is that advanced empathy expresses part of the client's experience that s/he is vaguely aware of, whereas confrontation points out experiences that are discrepant with or contradictory to the client's self-understanding. As such, confrontation has the potential to be both a more powerful and a more threatening response. Confrontation should occur infrequently, even less often than advanced empathy. On the basis of several studies, Hill and O'Brien (1999) concluded that, in psychotherapy, only about 1% to 5% of all counselor statements are confrontations. In genetic counseling you must be extremely careful when using confrontation. Trust building is essential before confronting clients. Confrontation may be most appropriate for predictive testing situations where you would see the client multiple times.

Guidelines for Effective Confrontation

When making a confrontation, you should attempt to be with rather than against your client (Leaman, 1978). We recommend the following strategies when using confrontation:

Formulate a Response

- Time your response. Use confrontation when your client is likely to be open to it. Direct confrontations at the very beginning have been found to be ineffective in consultation relationships (Dougherty et al, 1997). As we previously stated, rapport and trust must be developed before confrontations are likely to be heard. Stronger confrontations will only be successful if you and your client have a trusting relationship.

- Begin with accurate empathy. You must understand your client's experience before you can detect and raise issues of discrepancies or distortions.

- Moderate the depth. Decide how big a difference there is between what you want to say and what the client believes to be true (Pedersen and Ivey, 1993). If the difference is too big, your client will be more likely to reject your confrontation.

- Anticipate impact. Estimate your client's ability to handle the confrontation before you intervene. If your client seems to be confused or disorganized, you should wait until s/he is in a more receptive state.
- Use successive approximations. Introduce confrontation gradually; begin with small aspects that the client has some likelihood of being able to modify. Describe your client's behavior and its significance and/or consequences
- Choose your vocabulary and syntax carefully. Confrontation responses can sound accusatory or patronizing. You should speak tentatively ("I wonder if..."; "Perhaps..."; "Maybe..."; etc.), and use a questioning tone that leaves the client room to disagree.
- Check your motivation. Use confrontation to help the client, not in order to be right, to release your anger or impatience, to punish or get back at the client, or to put your client in her or his place. It is not appropriate to confront a client because you are bored, anxious, need to feel in control, or want to dominate the interaction.
- Be sincerely concerned. Communicate your confrontation in a way that demonstrates that you have a sincere interest in your client's welfare. Confrontation should be grounded in empathic understanding. For example, "You seem to have many different feelings about this; let me try to put them into words so we can look at what you're experiencing together" (Martin, 2000, p. 54). Furthermore, if your confrontations imply criticism, that is, if clients feel that you are accusing them or getting into a power struggle, your relationship and the session can quickly deteriorate (Martin, 2000).
- Use feedback skills. Since confrontation is a type of feedback, it is useful to consider guidelines for delivering effective feedback. As discussed in Chapter 1, Danish et al (1980) suggest that a good feedback giver:
— is focused on behavior rather than on the client's personal characteristics.
— gives only as much information as the client is ready to handle.
— makes the confrontation as soon as possible after the behavior has happened.
— is concise, tentative, and descriptive rather than judgmental, and only confronts about behavior that s/he believes the client can control or change. For instance, it is judgmental and does no good to tell a couple that they shouldn't have conceived in the first place.
— states the consequences of the behavior for the client.
— focuses on both strengths and weaknesses; asks the client to respond to the confrontation, and is willing to modify it based on the client's feedback.
— is definite (i.e., does not give the feedback and then take it back).

FOLLOW UP ON A CONFRONTATION RESPONSE

- Monitor the impact of your confrontation. Sometimes clients perceive your statements differently from the way you intended them (e.g., you may intend to point out a discrepancy in your client's story, while the client thinks that you are saying s/he is too confusing/inarticulate/stupid). To check out the impact of a confrontation, you could ask, "What do you think about what I just said?" or "How do you feel about what I just said?" (Leaman, 1978).

- Follow-up your confrontation. Confrontations can be threatening and painful to hear. You should follow up with supportive empathy statements that acknowledge your clients' experience. For example, "I know that this is hard for you. I can see why you try to cut me off when I'm telling you these painful things. Let's try to go more slowly, so that you can take this in gradually."

- Don't expect miracles. Not all confrontations produce insights that lead to change (Pedersen and Ivey, 1993).

Possible Behaviors to Confront

DISCREPANCIES IN INFORMATION

Confrontation of discrepancies is important in order to prevent confusion and to verify the accuracy of information (Pedersen and Ivey, 1993). This type of confrontation is common in genetic counseling because you must gather accurate data in order to help clients set goals and make decisions. Baker (1998) describes three types of information discrepancies:

- Gap: an issue usually associated with a particular genetic situation that is not raised.

- Omission: the client fails to include relevant information in her or his personal narrative.

- Inconsistency: between what the client says at different times in the session. For example, "Earlier you told me this is your first pregnancy, but now you mentioned a miscarriage."

DISCREPANCIES BETWEEN IDEAS AND ACTUAL BEHAVIOR

The prenatal client's partner says, "I only want what's best for my wife." You respond, "You say you only want what's best for her, but your opinions about options don't seem to coincide."

AMBIVALENCE

Ambivalence is a common human experience, and clients should be given permission to feel this way (Fine and Glasser, 1996). Example: "You say you want to have the testing done, but you keep canceling your appointment. I wonder if you have mixed feelings." Or the client says, "I'm only here because my doctor sent me. But since it took me hours to get here, and since I'm here already, I might as well have the amnio done." Counselor: "That's not a good enough reason. Let's consider the reasons you might and might not want the test."

DISCREPANCIES BETWEEN WHAT THE CLIENT SAYS AND THE REAL-WORLD CONTEXT (PEDERSEN AND IVEY, 1993)

For example, your client says, "My child is just a little developmentally delayed, but the doctors told me he'll catch up if we just work with him." You might respond, "You say that he's going to catch up, but your medical records indicate that he has Down syndrome."

DISCREPANCIES WITHIN THE CLIENT'S MESSAGES AND/OR INTERNAL DIALOGUE

"You've said that you could never have an abortion; you've also said that you couldn't deal with a child who had Down syndrome." Or "You've said you want to have an amniocentesis; you've also said that your life would be over if you had a miscarriage following the procedure."

DISCREPANCY BETWEEN CLIENT SELF-PERCEPTIONS AND GENETIC COUNSELOR PERCEPTIONS OF THE CLIENT

The client says, "I'll never be able to make a decision by myself!" You say, "And yet, you made the decision to come here and to have the testing done even though your family was against it. That suggests to me that you can be strong and decisive."

DISTORTIONS

"I'm wondering if blaming your child's condition on the way he was delivered keeps you from having to acknowledge your own medical history."

EVASIONS/AVOIDANCE

"You've told me that you've forgotten to ask your siblings to be tested. Is this perhaps because you know it would mean a more definite answer about your own cancer risk?" Or the client was supposed to request that his medical record be sent to the genetic counselor. Client: "I really didn't have a chance to call the doctor's office." Counselor: "I'm wondering if you really want to pursue testing."

NONVERBAL CONTRADICTIONS

The client says, with tears in her eyes, "I'm OK with the decision to terminate." You reply, "You say you are OK, but you look very sad."

GAMES, TRICKS, AND SMOKE SCREENS

"I wonder if by continually interrupting me, you are protecting yourself from hearing this painful information?" Or the client repeatedly says, "Yes, but. . ." Counselor: "You say 'Yes, but. . .' every time I suggest a resource; instead, you could investigate to learn more about Fragile X syndrome. I wonder if you really want to learn more about the condition."

CONFRONTATION FOR ENCOURAGEMENT

Client:　　"I can't think of anything I could say to my sister to persuade her that genetic testing is in both of our best interests." Counselor: "I think you've raised several persuasive points in talking with me."

Counselor:　"You say that you are too weak to handle the test results if they turn out to be positive. However, you seem strong and able to reach out to others for support."

Possible Client Reactions to Counselor Confrontation

Egan (1994) describes six ways that clients could respond to a confrontation:

- Deny the feedback. Your client may calmly tell you that your feedback is wrong, or angrily refuse to accept what you said.
- Discredit the source. For example, a common genetic counseling client response might be, "You don't understand. After all, you don't have Huntington's disease in your family."
- Try to change your mind. Your client might try to argue you into believing her or his behavior is really something else (e.g., "Oh, if you knew me better, you'd realize that I really can't handle this type of news!")
- Devalue the topic. For example, "My joking that our daughter's genetic condition is my husband's fault really doesn't mean anything!"
- Seek support elsewhere. "Well, all of my family members and friends agree with me!"
- Your client may pretend to agree with you.

Pedersen and Ivey (1993) identify several additional ways that clients could respond:

- Client is willing to admit to part of what you have confronted.
- Client agrees with the confrontation but refuses to do anything about it.
- The client chooses to compromise or accommodate the problem.
- The client hears the confrontation and uses the insight to change the behavior.

It is important to note that it may take time for clients to respond fully to confrontation. Their full reactions may not be evident in the genetic counseling session.

Challenges in Using Confrontation

Confrontation is not an easy intervention to make. We tend to avoid giving bad news to others because we fear their negative reactions, and/or it makes us feel bad or uncomfortable to think that we have caused someone else's pain: "Many therapists have said that this is one of the hardest aspects of doing therapy; it takes time and experience and personal strength to be able to go through intense emotional pain with another person who matters to you" (Martin, 2000, p. 59). Similarly, in genetic counseling you may avoid using confrontation with your clients because:

- You want to be liked and are afraid clients won't like you after you confront them.
- You don't want to hurt or embarrass clients (especially likely if you regard clients as fragile and vulnerable). "Kindness, at the expense of honesty, creates suffering and false impressions and distorts experiences of reality for oneself and others" (Wilbur and Wilbur, 1980, p. 130). They go on to say that, "Many counselors fail to realize that they do not directly cause pain through the honest expression of their perceptions of the client. Their honesty merely evokes the pain and conflict already present in the client" (p. 136).
- You have a cultural belief that confrontation is a rude or otherwise inappropriate behavior.
- You might be off-base, that is, you are afraid that you are biased against or wrong about the client.
- You might open yourself up to feedback from the client.
- Your client might get angry, shut down, or even get up and leave!
- You are afraid of sounding phony if you confront clients about their strengths (i.e., you are uncomfortable giving compliments).

Cultural Considerations in Using Confrontation

You cannot use confrontation in the same ways with clients from all cultural groups. You will need to be sensitive to cultural differences and modify your approach depending on a client's background. For instance, direct challenges with Asian, Latino, and indigenous American clients generally should be avoided (Ivey, 1994). Additionally, cultural practices for some Chinese individuals involve being extremely careful not to hurt another person or to make the person "lose face"; a direct confrontation would be viewed as very disrespectful, especially if the confronter was younger (Hill and O'Brien, 1999). Another implication of these cultural differences is that clients from some cultural groups might feel compelled to agree with your confrontations in order to prevent hurting you. Also, the appropriateness of confrontation between men and women differs across cultures (e.g., a female genetic counselor communicating with a male from the Middle East should be particularly careful about using this type of intervention).

Pedersen and Ivey (1993) recommend addressing the different rules that various cultures have about confrontation by:

- Being aware of your own cultural assumptions as well as those of your client's culture.
- Framing confrontations in ways that make it appropriate to your client's culture. Change the words or the process of communicating the confron-

tation; translate it into the client's cultural style so that your confrontation can be understood. For example, the use of the word *problem* might be ineffective for a client who comes from a culture where it is unacceptable to have a weakness.

- Trying not to be "distracted by behaviors—no matter how discrepant they might seem—until they are understood from the viewpoint of the client's values and expectations" (Pedersen and Ivey, 1993, p. 196). For instance, some black American clients will look away when listening to you.

Example: A Middle Eastern couple is counseled regarding prenatal testing. The husband does all of the talking but states that it is his wife's decision. The wife keeps her eyes on the floor and says nothing. Counselor: "Mr. __, you've said several times that it's up to your wife to decide about testing this pregnancy, yet she has said very little today. Help me understand how she will make this decision."

It is important to remember that confrontation is unlikely to work when clients hold strong cultural beliefs. In such situations, you need to respect their view and move on. For example, a couple from Pakistan who has a child with Friedreich ataxia does not believe that their consanguinity caused this condition. You could say, "I understand that you do not think your child's condition happened because you and your wife are related. Can we talk about some tests that would tell us if the next baby will have the same thing?"

It is important to be flexible in using confrontation with clients whose cultural backgrounds differ from yours. You generally should strive to make your confrontations gentle, since such feedback is difficult for most individuals to hear. Remember that it is always appropriate to ask clients to help you understand their cultural perspective on an issue. It may be sufficient to say, "You and I come from different cultures. Can you help me understand how we can approach this issue together?"

Closing Comments

Advanced empathy and confrontation are less frequent responses than other genetic counseling behaviors such as primary empathy, questioning, and information giving. Nevertheless, when used strategically, they can foster client insights about themselves and their situations. Often these insights will help clients achieve greater acceptance of their feelings, thoughts, and behaviors. They may also prompt changes in behaviors that are getting in the way of their goal-setting and decision-making processes. As a beginning genetic counselor, you may feel anxious about using these powerful responses. However, with supervised practice, you will gradually become more comfortable incorporating advanced empathy and confrontation skills into your counseling repertoire.

Class Activities

Activity 1: Class Discussion

Students talk about what they think advanced empathy and confrontation are, how the two skills are similar and different, and what functions they serve in genetic counseling. This discussion can be started by having students respond to these questions in dyads.

Estimated time: 10 to 15 minutes.

Activity 2: Subgroups

Each student selects and reads a client statement aloud (see Appendix 8.1), then discusses what it might be like to be this client. The student generates as many ideas as she or he can about:

- the client's surface feelings, thoughts, and issues
- the client's underlying feelings, thoughts, and issues.

Then the student speculates on what s/he might have to confront the client about during genetic counseling.

Students are NOT to give advice or solve the client's problem, but rather to focus on the client's feelings. Other group members can add ideas about the client's feelings, themes, etc.

Estimated time: 45 minutes.

Activity 3a: Low-Level Advanced Empathy and Confrontation Skills Model

The instructor and a volunteer genetic counseling client engage in a role-play in which the counselor demonstrates poor advanced empathy and confrontation behaviors (e.g., overanalyzes client motives and feelings; confronts the client in punitive, inaccurate, biased ways). Students observe and take notes of examples of poor advanced empathy and confrontation.

Estimated time: 10 minutes.

PROCESS

Students share their examples of poor advanced empathy and confrontation. Then they discuss the impact of the counselor's poor skills on the client. The client can offer her or his impressions of the counselor's behaviors after the other students have made their comments. The instructor could divide the group of students and have half of them focus on counselor behaviors and half focus on client behaviors. Or half of the class could focus on advanced empathy behaviors, and the other half could focus on confrontation behaviors.

Estimated time: 15 minutes.

- Students often have difficulty differentiating advanced empathy from confrontation. The instructor should assist them in categorizing the responses they observe in the role-play.
- One issue that may come up is that the counselor's intended response (e.g., advanced empathy) may be perceived differently by the client or observers (e.g., as a confrontation)
- One option is to pick up with a role-play used earlier in the class so that the counselor can more quickly move to advanced empathy and confrontation skills.

Activity 3b: High-Level Advanced Empathy and Confrontation Skills Model

The instructor and the same volunteer repeat the same role-play, only this time the counselor displays good advanced empathy and confrontation skills. Students take notes of examples of good advanced empathy and confrontation behaviors.

Estimated time: 15 minutes.

PROCESS

Students discuss their examples of good advanced empathy and confrontation and the impact of the counselor's behaviors on the client. They also contrast this role-play to the low-level role-play. The instructor could divide the group of students and have half of them focus on counselor behaviors and half focus on client behaviors. Or half of the class could focus on advanced empathy behaviors, and the other half could focus on confrontation behaviors.

Estimated time: 15 minutes.

INSTRUCTOR NOTE

- Students could work together in Think-Pair-Share dyads to identify advanced empathy and confrontation examples and their impact on the client.

Activity 4: Advanced Empathy and Confrontation Triad Exercise

Three students practice advanced empathy and confrontation skills in 15- to 20-minute role-plays taking turns as counselor, client, and observer. Allow another 15 to 20 minutes of feedback after each role-play. The students should focus on using good helping behaviors.

CRITERIA FOR EVALUATING COUNSELOR ADVANCED EMPATHY AND CONFRONTATION

Well-timed

Accurate

Tentative

Concise

Specific

Identifies themes, underlying issues, and/or discrepancies

Follow-up with primary empathy

PROCESS: IN THE LARGE GROUP DISCUSS:

- How was it to do this exercise?
- What are you learning about advanced empathy and confrontation in general?
- What are you learning about yourself?
- What questions do you still have about advanced empathy and confrontation?

Estimated time: 1 1/2 to 2 hours.

INSTRUCTOR NOTE

- The observer or counselor may wish to stop the role-play if the counselor is obviously "stuck." The triad can then engage in a brief conversation about the client's dynamics, issues, etc., in order to help the counselor. Then the counselor and client resume the role-play.

Written Exercises

Exercise 1: Journal Entry

Ask students to write a journal entry or short paper addressing the following:

1. Describe what you consider to be a confrontation.

2. Discuss how confrontations were/are handled in:
   - your family;
   - your peer group as a child;
   - your current peer group;
   - the cultural group with which you identify.

3. How might these experiences impact the way that you will approach confrontation with your genetic counseling clients?

4. In general, is it easier for you to make confrontations about a person's strengths? Limitations?

5. Are certain types of clients easier for you to confront?

Exercise 2: Primary Empathy, Advanced Empathy, and Confrontation

Read each of the following 14 client statements and write one primary empathy response, one advanced empathy response, and one confrontation response. Write your responses as if you were actually talking to the client.

[Hint: You may have to infer more knowledge about the client than is written here in order to formulate your advanced empathy and confrontation responses.]

For example, a 40-year-old man at risk for Huntington's disease, says, "I'm sick of worrying about this all the time! Every time I trip over something I think I have it. I think, 'Oh no, you're gonna end up just like your father.' Everyone made fun of him because of his disease. It was so hard being a child when your father acted so goofy. I don't know what I'll do if I find out I have this gene."

Primary empathy: "So you're scared that you have Huntington's, too?"

Advanced empathy: "Are you afraid your children will be saying the same thing about you?"

Confrontation: "This seems to be very distressing for you, and I'm wondering about the fact that you haven't had testing until now."

- 35-year-old prenatal client: "I'm afraid my husband will not understand my reasons for wanting to continue with this pregnancy. He might try to talk me out of my plan. I'm afraid he won't understand how I feel."

- 25-year-old male client: "I don't know whether to be tested for Huntington's disease or not. I'm so frustrated! You'd think after watching my father with this disease that I'd know what to do."

- 50-year-old woman talking about her 25-year-old child: "He knows that he can take advantage of me because of his albinism. If he gets sunburned or begins to talk about how he feels worthless and hopeless, I go crazy. He gets everything he wants out of me, and I know it's my own fault. But I still love him very much."

- 17-year-old male prenatal partner: "My girlfriend is pregnant, and the baby has abnormalities. She says she want to have an abortion. She says it's her problem and she can handle it without me. She never even asked me what I think she should do! I mean, it's my baby, too!"

- 10-year-old boy with Duchenne muscular dystrophy: "My classmates don't like me, and right now I don't like them! Why do they have to be so mean? They make fun of me because I can't walk or play with them. Gee, they don't have to like me, but I wish they'd stop making fun of me."

- 16-year-old with neurofibromatosis (NF) and a lot of visible nodules: "I'm only here because my mother made me come."

- 32-year-old mother and her husband who are from India and have three daughters: The mother says (looking tearful), "We want to have a boy."

- 38-year-old woman with a history of five miscarriages and no living children: "No one really knows how I feel. I know it's just miscarriages, and not like a real baby."

- Prenatal client who has a child with cystic fibrosis (CF): "My husband wants me to have prenatal testing for this pregnancy, but I just don't know. Susan, our daughter, is doing so well, and we love her so much."

- 30-year-old woman: "My family does not talk about how so many people in our family have cancer. I can't talk to my husband about it, either. And I don't know what to think about my two daughters!"

- Consanguineous couple from the Middle East: "I don't think this is a genetic problem. All of my sisters married cousins, and their children are all normal."

- 25-year-old African-American woman who has one child with sickle cell anemia, terminated a second pregnancy, is pregnant again and considering prenatal testing: "Everyone tells me that this baby will be affected because I ended the last pregnancy."

- 35-year-old Arabic woman whose newborn has Down syndrome: "What will she look like when she is older? Will she look abnormal?"

- 20-year-old Catholic, Hispanic prenatal client. Prenatal testing revealed abnormalities: "I'd like to have an abortion, but my priest and my husband are so against it."

INSTRUCTOR NOTE

- Students can be asked to also describe a cross-cultural counseling experience they were involved in or observed and to write three responses for this situation (i.e., primary empathy, advanced empathy, and confrontation).

Exercise 3: Role-Play

Engage in a 20- to 30-minute role-play of a genetic counseling session with a volunteer from outside the class. During the role-play focus on all of the helping skills you've learned so far. Try to include at least one advanced empathy and one confrontation response. Audiotape the role-play. Next transcribe the role-play and critique your work. Give the tape, transcript, and self-critique to the instructor who will provide feedback. Use the following method for transcribing the session:

Counselor	Client	Self-Critique	Instructor
Key phrases of dialogue	Key phrases	Comment on your own response	Provides feedback on your responses

Annotated Bibliography

Fontaine, J. H., Hammond, N. L. (1994). Twenty counseling maxims. Journal of Counseling and Development, 73, 223-226.
[Provides 20 specific suggestions to help beginning counselors transition from the classroom to the clinical setting. Several of their recommendations are useful guidelines for advanced empathy and confrontation.]

Handelman, L., Menahem, S., Eisenbruch, M. (1989). Transcultural understanding of a hereditary disorder: Mucopolysaccharidosis VI in a Vietnamese family. Clinical Pediatrics, 28, 470-473.
[Discusses the importance of ascertaining patients' understanding of their condition, particularly their culturally based nonmedical beliefs.]

Kessler, S. (1997). Psychological aspects of genetic counseling. X. Advanced counseling techniques. Journal of Genetic Counseling, 6, 379-392.
[Provides examples of positive confrontations.]

Kessler, S. (1999). Psychological aspects of genetic counseling. XIII. Empathy and decency. Journal of Genetic Counseling, 8, 333-344.
[Defines empathy and offers specific clinical examples.]

Leaman, D. R. (1978). Confrontation in counseling. Personnel and Guidance Journal, 630-633.
[Defines confrontation and its function in mental health counseling, recommends guidelines for its effective use, and provides several examples.]

Ottens, A. J., Shank, G. D., Long, R. J. (1995). The role of abductive logic in understanding and using advanced empathy. Counselor Education and Supervision, 34, 199-211.
[Describes an information processing approach for formulating hypotheses about clients. It is intended to help novice counselors make sense of client data in order to gain an advanced empathic understanding of the client.]

Wilbur, M. P., Wilbur, J. R. (1980). Honesty: expanding skills beyond professional roles. Journal of Humanistic Education and Development, 24, 130-143.
[Presents several arguments for the importance of being honest in relationships with clients and the many difficulties inherent in taking an honest stance. Honesty includes both confrontation and advanced empathy behaviors.]

Appendix 8.1: Client Statements for Advanced Empathy and Confrontation (Activity 2)

- "I've just found out that my brother is gay. I already know that two of my cousins on my dad's side are gay. I've read that homosexuality may be inherited. Are my wife and I at a greater risk of having a gay child if she ever gets pregnant?"

- "I'm 39 and this is my first pregnancy. I've just been told that the amnio showed something wrong with my baby. They say there's a chance my child could have a lot of behavior problems in school and may not be able to learn. Oh, and I remember they also said he would be really tall. But they aren't real sure because hardly any research has been done on people with this condition. What should I do?"

- "I asked to see you first before my husband comes in because I'm not sure I want to go through with this amniocentesis. He knows I was pretty wild in my younger days, and I did a lot of partying, including smoking pot and dropping acid a few times. What if the amnio results come back and there's something wrong with the baby?"

- "I have just tested positive for Huntington's disease. I'm not symptomatic yet, but I want to know exactly what to expect. How do other people deal with this? Have you ever heard of people with Huntington's committing suicide?"

- "My husband and I just learned I'm pregnant. I believe my health is good and there are no genetic problems in my family. However, my husband is an active alcoholic. I am concerned about fetal alcohol syndrome."

- "I came here today for prenatal testing. This is my second pregnancy. My first child suffers from multiple genetic abnormalities, which will mean that he will be completely dependent on my husband and me for the rest of our lives. I am afraid that this might happen again, but I don't believe in having an abortion. I also want this second child to make sure the first child wasn't my fault."

- "This is my fourth pregnancy. I have three sons. My husband and I really want a girl this time. We have heard that some prenatal testing could determine the sex of the child. I would like to get this testing done here."

CHAPTER 9

Client Issues: Resistance, Client Affect, and Client Styles

Learning Objectives

1. Define resistance.
2. Identify counseling responses for addressing client resistance.
3. Describe selected types of client affect and their impact on the genetic counseling relationship.
4. Describe client stylistic differences and their effects on the genetic counseling session.

This chapter discusses three types of client characteristics that impact the process and outcome of genetic counseling: resistance, defense mechanisms, and emotional reactions. Your challenge as a genetic counselor is to recognize and address these factors using the basic counseling skills presented in this manual. This chapter also describes client stylistic differences and how they may play out during genetic counseling.

Client Resistance

Definitions of Resistance

Generally speaking, resistance is an individual's opposition to a process and/or outcome. Within genetic counseling, resistance refers to client behaviors that impede the work of genetic counseling. It is often an unconscious process that indicates a client is not fully committed to the genetic counseling relationship/session.

Some authors distinguish between reluctant clients and resistant clients (cf. Harris and Watkins, 1987). Reluctant clients do not want to come for genetic counseling at all, and you probably will not be able to counsel them effectively until you have first addressed their reluctance. (For example, if they state that they are there only because their primary care provider told them to come, you might say, "Is there anything that you feel you might gain by being here?") Then you could suggest some possible benefits in addition to any that your clients identify.) Resistant clients, on the other hand, may originally have been reluctant, but now they are more or less willing to attend counseling. However, they may still be unwilling to participate fully during the genetic counseling session.

Causes of Resistance

Harris and Watkins (1987) identified several reasons for client resistance in mental health counseling that we have modified for genetic counseling. Genetic counseling clients may be resistant because they are afraid or resentful, because they misunderstand, or because they are disconnected from the counselor. These reactions may be due to one or more of the following reasons:

FEAR

- Are afraid of taking personal responsibility for their situation and resulting decisions (e.g., client who always lived as if she had the Huntington disease gene and did not want to find out that she may not have the gene).
- Feel demoralized because of the genetic condition and also feel as if their privacy and independence have been threatened.
- Feel a loss of control and use resistance as a way to hold on to some power and self-esteem. Also, they may be ambivalent about making a decision that has important benefits but also significant restrictions (e.g., having genetic testing for breast cancer to better plan monitoring for the disease, but not wanting to hear that they have the gene).
- Have already experienced a great deal of emotional pain such as grief, anger, fear, shame, or guilt from their genetic concerns and are reluctant to discuss these feelings further (e.g., they may be ambivalent, wanting answers to their questions, but feeling afraid of what those answers will be).
- Believe that discussing their genetic concerns will bring into the open weaknesses or failures; thus, their self-esteem is threatened.
- Fear the unfamiliar (e.g., do not know what genetic services involve; feel apprehensive about discussing personal matters with a stranger, etc.).
- Are afraid that the genetic counselor will tell others what they said.
- Are testing the genetic counselor's level of support and competence.
- Feel threatened (e.g., fear being overwhelmed by a diagnosis of a genetic condition).

RESENTMENT

- Feel coerced (e.g., client was told that she could only have an amniocentesis procedure if she first spoke with a genetic counselor).
- Feel angry about being referred by others (family doctor, infertility specialist, etc.) and carry this resentment over to the genetic counselor.
- See no reason to go to a genetic counselor in the first place.

- Have a negative attitude about medical agencies, are suspicious of medical personnel, etc.

MISUNDERSTANDING

- Do not know how to participate effectively in genetic counseling because they do not adequately understand what genetic counseling entails.
- Believe that genetic counseling is essentially the same thing as psychotherapy, and therefore that it may involve extensive probing into their innermost experiences and motivations.
- Have unrealistic expectations about what genetic counseling can offer and become resistant upon realizing that their expectations cannot be met (e.g., the counselor cannot guarantee a healthy baby).
- View the genetic counselor's goals as different from their own (e.g., a client who believes that the counselor's agenda is to persuade her to have presymptomatic testing when she does not wish to do so).

DISCONNECTION WITH THE COUNSELOR

- Dislike something about the genetic counselor but do not bring up this negative reaction.
- Are culturally distant from the counselor with respect to gender, ethnicity, social class, religion, previous experiences with prejudice and discrimination, etc.

Behaviors That May Indicate Resistance

Resistance may be occurring when clients:

- Do not seem to know what they want from genetic counseling.
- Present themselves as not needing any help.
- Are only there because of someone else's urging, referral, etc.
- Express resentment about being there.
- Talk only about safe or low-priority issues (e.g., focus on the risk numbers and not on the condition itself).
- Are directly or indirectly uncooperative (e.g., refusing to discuss certain issues or *selectively forgetting* important aspects of their family history).
- Unwarrantedly blame others for their situation.
- Show no willingness to establish a working relationship with the genetic counselor.
- Are slow to accept responsibility for the decisions they need to make.
- Are irritable or actually hostile toward the genetic counselor.
- Seek to get support for a decision they have already made rather than being open to exploring and engaging in a decision-making process.
- Use one or more defense mechanisms (discussed later in this chapter).

Responding Effectively to Client Resistance

There are several strategies you might use to address client resistance:

- Explore the reasons. Remember that clients have no obligation to attend genetic counseling, and part of your job is to determine whether they want counseling.

EXAMPLES

If a client seems rushed or in a hurry, acknowledges this; the client may truly have somewhere else that s/he needs to be and will be very appreciative that you took the time to notice and understand this need. Clarifying that a client is indeed in a hurry will also alert you to be as efficient as possible in your use of session time.

Find out if financial coverage is a concern. Clients may be resistant or afraid because they fear having to pay out-of-pocket for yet another service.

A client was referred for genetic counseling to discuss her risk for having a baby with tuberous sclerosis (TS). She recently had a miscarriage. The client received her diagnosis 3 years ago when her previous pregnancy was diagnosed with cardiac tumors. The baby and mother were examined upon delivery, and both were given a diagnosis of TS. The client never returned to the medical geneticist, and a cardiologist had followed only her son. When the counselor was assessing the client's understanding of why she was referred, the client stated that she had no idea. The counselor explored this further with her, and the client admitted that the cardiologist wanted her to come in to discuss TS and her risk for having another baby with TS. Further exploration revealed that she did not believe that she had TS, that her findings of TS were familial birthmarks. Because she was in denial regarding her diagnosis, the genetic counselor focused on her feelings about her diagnosis, why she did not believe it, etc. The client was able to engage in a dialogue about these issues, feelings of guilt that she passed this on to her son, her fear of passing the condition on to other children, etc.

A client came in and sat down with a *huff* and crossed her arms. The counselor said, "You seem uncomfortable. Is there something I can do?" The counselor hoped this response would allow the client to vent her anger, and the counselor could determine the reasons for her anger.

The genetic counselor's role is to briefly consult with clients in the craniofacial clinic to determine the need for a complete evaluation in pediatric genetics. Some of the clients have multisystem involvement (not just craniofacial abnormalities) and consequently have been seen by multiple specialists. When the genetic counselor recommends yet another evaluation, she is more often than not met by some amount of resistance. In most cases it relates to a lack of understanding and/or a view that the counselor's goal of evaluation is different from that of the family's (e.g., the family mistakenly believes it is *to provide a label* rather than to insure quality care).

• View some degree of resistance as natural and normative (usual).

Example of a Partially Successful Attempt to Address Client Resistance and Hostility

The client was referred to one of the agency satellite clinics for genetic counseling due to advanced maternal age. She was extremely hostile right at the start. She was unhappy with the appointment time, felt that staff should be available in the evening since the hospital is open 24 hours, was unhappy that the staff was available only on Tuesday (but was unwilling to go anywhere else that staff would be available), and was unhappy that there was an hour's delay before her amnio was scheduled. The client proceeded to instruct the genetic counselor to "fix the problem." The genetic counselor tried to make a contract, to empathize (e.g., fear of unknown, anxiety of being offered an amniocentesis, fear of having an abnormal baby, etc.). She empathized with the client's frustration regarding appointment time, etc. The counselor tried advanced empathy (see Chapter 8), stating that the client's anger may stem from her fear; and she tried to empower the client by asking what information the client would like to hear. The counselor learned that the client was upset about having recently received a diagnosis of a thyroid problem. The client stated that she has been extremely stressed and hostile to everyone. She was also upset that no one told her not to go back to work after the amnio procedure, and she could not move beyond this issue. The genetic counselor tried to problem-solve with her, but the client found a reason why every suggestion would not work. The client never addressed the underlying issue. She accused the genetic counselor of not understanding, but would not help the counselor to understand.

• Regard resistance as clients' avoidance of what they consider to be a frightening and/or unhelpful experience. Discuss concretely the potential benefits of genetic counseling. We recommend doing this at the beginning of the session as part of the informed consent process.

Examples

The client was referred for genetic counseling late in the pregnancy (24 weeks). She had an abnormal triple screen 2 months ago and did not get referred until just recently. The client was frustrated with her referring physician and angry that she now had limited options if there was a problem with the baby. The genetic counselor let her know that she understood why she was frustrated and acknowledged that the client was in a difficult position. The counselor's empathy allowed some alignment and also gave the client time to express her anger, which helped her move forward in the session.

A client came in for a consult prior to an ultrasound for advanced maternal age and said, "I don't see why I have to have a *consult*. I'm not having any testing." The counselor explained that a consult was being done so that the client could understand the limits of the ultrasound test, and so that if problems were identified, the client would understand her options and could make an informed decision about those options.

Often families have the impression that the evaluation will involve genetic testing, which they may perceive (for whatever reason) as a negative experience. When it is explained that the process will involve gathering information, obtaining a history, doing a physical exam, and then considering or discussing testing if applicable, their resistance often disappears.

- Examine the quality of your own responses. Ask yourself if you are doing anything to generate resistance from your client.

Example

The client was a young woman with Turner syndrome who was extremely anxious. She had numerous misconceptions about genetic counseling. The counselor realized toward the end of the session that the client had imagined all sorts of horrible procedures that were going to be done to her. The counselor told her they would not be doing any tests that day, and the client immediately calmed down and began to listen to the counselor. Picking up on this earlier in the session would have allowed the counselor to address the client's anxiety and misperceptions even sooner.

- Accept and work with the client's resistance. Start with your client's frame of reference. Let your client know that you understand how s/he might feel. Accept your client's right to think differently, as this helps the client to maintain dignity (Larrabee, 1982). Acknowledge that your client is in a difficult situation and may have limited options (Hanna et al, 1999).

Example

A client came for an amnio and said, "I don't want to hear any numbers or anything that is scary." The counselor responded, "Genetic testing is very scary; sometimes we tend to give too many numbers in the hopes of helping clients make informed decisions. What kind of information can I give you that may make you more comfortable with this situation?"

- Invite client participation in every step of the process. Share your expectations and discuss your clients' reactions to being in genetic counseling. Give clients as much power as possible; focus first on their reasons for being there (Harris and Watkins, 1987). Ask them what you can do to help make their situation/decision easier.

Examples

A client came in and stated that she didn't know her due date. Her physician told her to come to find out about the effects of her medications on the pregnancy, but since she didn't know her due date, she didn't know what medications she actually took. The genetic counselor explained that it was very important to know her due date before talking about medications and was able to rearrange her schedule to get the client in for an ultrasound prior to the consult.

Sometimes it is apparent that the client has a need to be in control of what happens during the session. It is sometimes possible to accommodate this type of client depending, of course, on the circumstances. For instance, a genetic counselor was counseling the mother of a boy with developmental delay. The mother said that she wanted an appointment for Fragile X syndrome testing only. She was initially reluctant to agree to a complete evaluation, but once it was made clear that the Fragile X test would be done and that the other components were necessary for completeness, she became a willing participant. She had also initially refused to consent to follow-up counseling if the results were positive, but at the end of the session she agreed to it, in part because she felt that she had maintained some control.

- Help clients see how their resistance is impeding the genetic counseling relationship, if it is. Advanced empathy and confrontation (Chapter 8) may be helpful responses.

Example

A client came to learn more about her chance of having a baby with a birth defect. Her sister had a baby that died, and she did not know the cause. She refused to talk to her sister even after the genetic counselor explained that she needed the baby's medical records to determine if there was any familial risk. However, the client kept calling the counselor to ask if there was any prenatal test or other kind of test she could have. The genetic counselor finally needed to tell the client that she could not help unless the client talked to her sister.

- Search for incentives for moving beyond resistance, but use these sparingly.

Example

A 38-year-old woman who was 18 weeks into her first pregnancy came for genetic counseling because her triple screen showed her risk for Down syndrome was 1 in 50. Her ultrasound showed that her dates were correct. She could not decide whether to have an amnio. She was insistent on having her triple screen repeated despite the genetic counselor's telling her that this would not provide her with useful information. The genetic counselor steered the conversation toward a discussion about what the client would do should she find that her baby had Down syndrome. The counselor pointed out that if the client would consider a termination, then repeating the triple screen would not be a realistic option because of time constraints.

- Do not take client resistance personally.

McCarthy Veach and her colleagues (2001) found that a major challenge for genetic counselors and primary care providers was addressing diversity issues. In their focus group study, they heard accounts of Mexican immigrants who believed that if they discussed the possibility of a genetic condition with a health care provider, that the provider would "give" the condition to them. It was important for the health care providers to recognize that this resistance was not due to their own competency or personality. Rather, it resulted from a deeply held cultural conviction.

• Avoid getting into power struggles, which you will almost never win (Hanna et al, 1999).

After explaining the genetic counseling process, offer the client the option of declining.

It can be very frustrating when clients don't listen to what you are saying. You may find yourself becoming visibly agitated. Periodically, tune in to your body language, especially when you feel that your "buttons" are being pushed. First, you need to know what your buttons are (see Chapter 12 on counter-transference), then you need to practice checking on your reactions during the session. Finally, you need to step back and ask yourself why the client is not listening before deciding how to respond.

Other Strategies

Harris and Watkins (1987) recommend these specific strategies for reluctant and resistant clients:

• Emotive: Reflect client feelings about being coerced (e.g., resentment, anger, etc.)

• Cognitive: Challenge client beliefs that they have no choice to be there and no choice over their decisions about their situations.

• Definitional: Clarify your role and the client's role to demonstrate how the client has some choice and to separate yourself from coercion. Attempt to set mutually acceptable goals.

• Provide structure (e.g., informed consent) to clarify genetic counseling and to alleviate some anxiety.

• Use less threatening terms if your client is reactive, for example, "consultation" or "discussion" rather than "counseling"; "we are going to talk about some available options" rather than "testing procedures"; "testing options in pregnancy" rather than "prenatal diagnosis"; "changed" or "altered" gene rather than "mutation." Listen to the words your clients use to determine acceptable terms.

Defense Mechanisms and Coping Behaviors

Definition of Defense Mechanisms

Defense mechanisms are unconscious responses to a real or perceived threat (Clark, 1991). They are an attempt to maintain some measure of personal control and esteem and to reduce painful emotions in the face of this threat (Clark, 1991). Defense mechanisms temporarily protect an individual from anxiety, grief, guilt, shame, remorse, embarrassment, fear, and other painful feelings, and allow an individual to continue believing that the world is the way s/he thinks it is or wishes it to be (Hultman, 1976; Liburd, 1978). Defense mechanisms are used to some extent by all individuals, including counselors (Liburd, 1978)!

There are several different types of defense mechanisms that we define and illustrate with examples in Table 9.1.

Definition of Coping Styles

All defense mechanisms are attempts to cope, but not all coping strategies are defense mechanisms. Djurdjinovic (1998) identifies eight coping styles, which she defines as strategies for solving problems or for modifying the meaning of an experience. Three of these coping styles—seeking social support, plan, and positive reappraisal—have the greatest likelihood of leading to positive outcomes for clients. The other five are similar to the defense mechanisms discussed earlier, and they may be problematic if engaged in intensely and for very long periods of time. (Remember, defenses offer only *temporary* protection from reality.)

- Confrontative: tries to change the opinions of the person who is in charge (e.g., genetic counselor, physician, etc.).
- Distancing: acts as if nothing has happened (e.g., the client leaves the session after learning that he has a very high risk for having the gene for a form of colon cancer and therefore that his children are at risk, and says, "Well, I really didn't learn anything new today").
- Self-controlling: keeps feelings to oneself (e.g., when asked how she is feeling after receiving abnormal test results, the client says, "I'm fine" and says nothing more).
- Seek social support: talks with others in the hope of learning more (e.g., attending support groups for people with similar genetic conditions).
- Accept responsibility: criticizes or blames oneself.
- Escape-avoidance: hopes for a miracle.
- Plan: identifies next steps and follows-through on them.
- Positive reappraisal: tries to see any possible positive results or outcomes (e.g., "Well, I'd rather have a child with Down syndrome than have no children at all.").

TABLE 9.1. Client defense mechanisms

Definition	Client examples
Denial: Rejecting the possibility that an event happened.	"Nobody ever told me I was at risk." "He'll grow out of it." "She looks just like my mother."
Displacement: Shifting response from original aim to a vulnerable target.	"I won't get any useful information from your incompetent lab."
Identification: Assuming the attitude or behavior of an idealized person or group.	"My sister says it's foolish to worry about a few ultrasound *glitches*." "My friend says if I have an amnio that I'll have a miscarriage."
Intellectualization: Avoiding intolerable feelings through abstract, precise thinking with little or no feeling.	"So, it is a statistical probability that with this type of translocation a fetus would expire."
Projection: Blaming other people or situations for difficulties the client experiences.	"I know you think I'm a fool to continue this pregnancy." (In reality, the *client* feels foolish.)
Rationalization: Justifying objectionable information with plausible statements.	"Everyone has some *abnormal* genes... I'm probably not any more at risk than anyone else." "But I take my prenatal vitamins every day!"
Regression: Reverting to developmentally less mature behavior.	A well-educated and articulate couple, upon hearing abnormal test results, suddenly was unable to process any further information. They kept saying, "What do you mean?"
Repression: Putting intolerable thoughts and feelings out of one's mind.	"I don't recall having any siblings who died at birth."
Undoing: Canceling-out a distressing experience through a reverse action.	An obsessive need for prenatal testing after having a child with Down syndrome.

Adapted from Clark (1991).

Here is an example of addressing client resistance and client use of a positive reappraisal: A genetic counselor had an emergency session with a woman who had spastic cerebral palsy and who was pregnant with her third child. She was in a wheelchair, and although she appeared slightly retarded she was actually quite sharp. It was clear from the beginning that she did not want to be there, and when the counselor asked her why she came, she responded, "My doctor made me come down here because he thinks it would be a tragedy if I had a child just like me. He thinks I would want an abortion if the chances are high. Well, my mother gave me a chance at life, and I intend to do the same for my child. I'm doing just fine." By the end of the session, it was clear that she had come not for the counseling but to make a point. In this situation the genetic counselor effectively attempted to see the issue from the client's perspective. She was willing to hear her out, established a working relationship, and allowed the client to leave with the understanding that genetic counselors are not gatekeepers to abortion, but individuals trained to help clients meet the challenges presented by any given scenario/diagnosis.

Addressing Client Defenses

Clients need coping strategies to deal with intense experiences, and some clients may need to engage in a certain amount of defending before they can move on to more positive copies strategies that support effective problem solving and decision making. If you decide to address client defenses, you must first establish a trusting relationship. Begin by trying to understand your client's perspective via primary empathy and questions. Then indirectly point out defenses by commenting on client inconsistencies between verbal and nonverbal behaviors, contradictions, omissions, and misinformation (Clark, 1991). (See Chapter 8 on advanced empathy and confrontation for examples of possible counselor statements.) You should be careful about confronting defenses *head on* as that may actually increase their intensity and decrease client trust (Clark, 1991).

Client Affect

There are several different types of client emotions. These feelings are central to many genetic conditions/situations, and you should address them in genetic counseling. Direct expression of feelings may reduce their intensity for clients and may result in more effective problem solving and decision making (Hendrick, 1977). To help clients express their feelings, you should be accepting, encouraging, and invite them to fully describe what they are experiencing.

Client Anger

Anger is a complex emotional reaction that may mask deeper feelings such as fear and despair (Djurdjinovic, 1998). Anger may be an expression of frustration about one's situation (e.g., that a diagnosis has not been possible, or encountering one obstacle after another when attempting to obtain appropriate health care), or they may mask feelings of vulnerability (e.g., extreme fear about the implications of a genetic condition). Clients may be more likely to feel anger when a genetic condition is uncommon in their family and respond with statements such as, "Why did this happen to me?" or "What did I do to deserve this?" (Djurdjinovic, 1998; Murray, 1976). The more important or meaningful an issue is for a client, the more likely the client is to be angry (Gintner, 1996). Positive aspects of anger include that it gives clients emotional energy to deal with problems and it can "clear the air," allowing clients to release pent-up feelings (Geldard, 1989). Visible clues that a client may be angry include breathing more rapidly, sweating, clenching muscles (for instance, in the jaw or the hands), becoming flushed, and raising one's voice.

Anger can be a difficult emotion to address, especially for beginning counselors (Geldard, 1989). Nevertheless, it is important that you do so because anger will impede the work of genetic counseling if not expressed. "Counselors who consciously or unconsciously avoid situations that will result in being targeted are like surgeons who want to operate successfully without getting blood on themselves" (Cavanagh, 1990). You must be prepared to be the target of client anger and should respond nondefensively.

First, it is important to realize that you are not really the target (Djurdjinovic, 1998); the client may be displacing anger (for instance, about abnormal test results) onto you. However, you may want to check with the client to see if you have perhaps done something to elicit anger. For example, McCarthy Veach and her colleagues (1999) found that some clients resented the genetic counselor's discussing abortion after they specifically stated that it was not an option they would consider. Next, acknowledge *inside* that you may feel like retaliating, but do not do so. Instead, begin with basic primary empathy: "I can see that you are very angry right now." This type of reflection indicates that you respect the client's feelings. Finally, you should talk about what is making your client angry: "Can you tell me what you're angry about?"

Grief

Grief occurs when individuals lose something important to them (e.g., the perfect baby, the loss of one's health) (Djurdjinovic, 1998; Murray, 1976). Grief is universal because we all experience losses in our lives, and we all go through a recovery process (Hopkins, 1997). Often grieving individuals will cry. During the genetic counseling session, many clients will be on the verge of tears, although some may attempt to hold back their tears (out of embarrassment, fear of appearing weak, etc.). You should give them permission and provide an accepting atmosphere in which to cry. Both verbally and nonverbally show that you will listen with concern: "It's OK to go ahead and cry." Put a box of tissues within easy reach. These actions suggest that you are comfortable with clients who lose their composure.

It is also important to reassure clients that what they are feeling is a normal reaction to loss and that grieving takes some time (Geldard, 1989). Experts estimate that the grieving process can take from 6 months to 5 years. To reassure clients and to normalize their experience, you might say, "You've lost so much. I can see how you would feel so much pain." After your client responds to your statement, you could follow up with a statement such as, "Grief takes a long time, so don't feel you have to get through it quickly."

It is important to let clients discuss their grief, even if it is a story that they have told before (Geldard, 1989). It may be helpful to let clients repeatedly discuss their loss because grief is a repetitive process (Murray, 1976). Although clients experience and express their grief in highly individualized ways and will vary in the sequence and timing of their grief process, the following aspects of grief are fairly common (Geldard, 1989; Ormerod and Huebner, 1988):

- Shock, especially when the loss was unanticipated. The client may have a great deal of difficulty accepting the meaning of a diagnosis (Ormerod and Huebner, 1988).

- Denial that it really happened. Denial may allow clients to hold on to hope until they are able to come to terms with the situation (Ormerod and Huebner, 1988). Genetic conditions can be so traumatic that individuals will deny their existence. For example, unless an infant has obvious physical abnormalities, the parents may act as if the child is fine (Murray, 1976). A possible counselor response in this situation is: "I get the impression that you think your baby will be fine. I also heard you say that your doctor thinks your baby has Down syndrome. Do you think your doctor might be right?" This allows your client the opportunity to either stay in denial or move forward. You might suggest that parents put off reproductive decisions while they are in denial, as denial usually will dissipate over time (Murray, 1976).

- Physical and psychological symptoms (e.g., insomnia, loss of appetite, depression, hopelessness).

- Guilt for passing along a defective gene to their child. Guilt and anger are particularly likely when the etiology of a condition is unknown (Ormerod and Huebner, 1988).

- Anger at the medical professionals for not arriving at a diagnosis sooner, for not being sympathetic enough, etc.; or anger toward the person they lost— "If she had treated her breast cancer sooner, she might have lived"; or anger at God/higher power, which can be particularly problematic for religious clients who believe it is not OK to be angry with God. In this latter situation, you might explain that anger is a normal part of the grieving process and suggest that the clients look for support from their religious community.

- Idealization of the person or thing they have lost (e.g., believing that their lives would have been perfect if their baby had lived).

- Realism that the loss is permanent. At this phase and in the subsequent acceptance phase, clients will be more capable of hearing and understanding information and they will be better able to evaluate the information that they receive (Ormerod and Huebner, 1988).

- Acceptance of the loss. Until your clients can accept their situation, you must show patience and support and allow them to be indecisive, confused, angry, etc. (Ormerod and Huebner, 1988).

- Readjustment

- Personal growth

It is a mistake to try to falsely reassure grieving clients that everything will be fine or to try to cheer them up (Geldard, 1989). You must let them freely express their emotions, some of which may be quite intense. One problem, however, is that beginning counselors often are uncomfortable with intense client emotion. If this is the case for you, acknowledge your discomfort internally and later discuss it with your supervisor or a colleague.

When genetic counseling involves prenatal or perinatal loss, it is important to attend to the father's grief as well as to the mother's. The literature emphasizes maternal attachment and mourning, which has led to fathers sometimes feeling marginalized, invisible, and as if they are responsible for their partner's well-being (Rich, 1999).

Anxiety

Clients are frequently fearful and anxious about genetic counseling (Kessler, 1992a,b; McCarthy Veach et al, 1999; Murray, 1976). If their anxiety is too intense, it will disrupt their thoughts and behaviors. For example, they may have difficulty comprehending and retaining genetic and medical information. Clients will not typically express their anxiety directly. Instead they may show it indirectly (e.g., repeating the same question, avoiding sensitive topics by changing the subject, joking, making trivial comments, frequently interrupting you, seeking excessive reassurance from you, behaving dependently). Anxiety is a very contagious emotion. Often you will find yourself beginning to feel uneasy; and this can be a clue that your client is anxious.

First you must recognize your client's anxiety. Next you should remain calm (take a couple of deep breaths); your calm demeanor may have a calming effect on your client. Finally, you should reflect your client's anxiety: "You seem to be nervous. Can you tell me what's making you feel that way?" Talking about your client's anxiety in a calm, accepting way may help to diffuse it.

Guilt and Shame

Guilt and shame are common emotions, especially when clients are the parents of children who have genetic conditions (Djurdjinovic, 1998; Weil, 2000). Some clients feel guilty for having negative feelings about a child or family member with a genetic condition (Ormerod and Huebner, 1988). Other clients feel guilt because they believe they have done something wrong and are being punished by God or a higher power (Murray, 1976). It can be very difficult to discuss the scientific basis of genetic situations with clients who view their situation as punishment for their own or their ancestors' sins. Until some of your client's guilt is addressed, you probably will not be very successful at communicating genetic information.

Consider the following example: The genetic counselor was counseling a Korean couple who shared with her their belief about why their son had muscular dystrophy. They had an ultrasound study during the pregnancy that predicted that the fetus was female. But the wife's father had prayed for a boy. They felt that this "conversion" from a girl to a boy resulted in the mutation causing the disorder in their son, and thus they felt responsible (guilt) for having caused the disorder. The genetic counselor certainly could not dispel their belief, but she felt that the couple's awareness and their willingness to share their belief with her spoke volumes about their progress toward acceptance of this diagnosis.

"Shame is a feeling of being flawed and defective as a human being" (Loughead, 1992, p. 127). Murray (1976) writes about the *psychology of defectiveness* and says that any clients who receive a genetic diagnosis may feel guilt and shame about being less than perfect and perceive themselves as socially unacceptable and unworthy: "If I have a defective gene, then I must be all bad." This irrational feeling may also be socially based, especially when families with one or more children with genetic conditions have already been branded by the community as defective (Murray, 1976).

Shame is a difficult feeling to acknowledge. Clients who feel shameful may engage in repression by keeping awareness of their shame out of conscious experience and by refusing to think about the situation that has led to this feeling (e.g., a diagnosis of a genetic condition). Emotional clues that a client may feel shame include chronic low levels of depression, uneasiness or anxiety, and guilt (Loughead, 1992). Behavioral clues include shaming others ("It's your fault that we have a retarded child!"); being critical or blaming ("This whole process wouldn't be so awful if you had been clearer about how severely retarded our child would be"); focusing outside of oneself (talking about everyone else's feelings and reactions except one's own; keep in mind, however, that in some cultures *externalizing* is a typical behavioral pattern.); forgetting or lying about critical information; and overcontrolling another person (e.g., the overprotective parent of a retarded child).

As with other emotions discussed in this chapter, unexpressed guilt and shame impede the genetic counseling process. If you have developed rapport with clients, you may be able to invite a discussion of their guilt and shame by using primary empathy and by maintaining an accepting, nonjudgmental attitude.

You might also try addressing guilt and shame by using advanced empathy to reframe the issue for clients. This may help them see things from a different perspective (Kessler, 1997b). For example, parents whose child has Down syndrome were struggling with guilt and shame because they came into the session believing that they had caused this to happen. The genetic counselor explained the simple mechanics of meiosis and emphasized that it was no one's fault—no one can make it happen or prevent it from happening. She reinforced this message by saying that this is a simple error in the division of the genetic material, which can happen to anyone.

Client Styles

No two clients will be exactly alike. They will show stylistic differences. A *style* suggests a certain degree of consistency in the way a given individual behaves across different situations. For instance, a person who expresses emotions easily and intensely may be very emotional in most situations, including genetic counseling. On the other hand, individuals who typically are controlled would be less likely to express strong emotions during genetic counseling. Client stylistic differences influence how they respond during genetic counseling sessions. Additionally, counselors have a preferred intellectual and emotional style that affect how they approach their clients. To the extent possible, counselors should vary their approach to effectively work with different types of client styles.

Cheston (1991) identified client styles for two major dimensions: intellectual and emotional. She also described the importance of the spiritual dimension, although she did not articulate any spiritual styles. We have included the spiritual dimension because it is important and variable.

Intellectual Styles

- Inductive reasoners gather a large amount of data and then make generalizations based on these data. Inductive reasoners may display a fair amount of confusion until they can find the patterns in the data (i.e., draw a conclusion from the facts). Then they may have "light bulb" experiences as the data suddenly fall into place for them. Inductive reasoners may also provide you with a great deal of detail. You must not allow yourself to get bogged down in these details. The most effective counseling strategy is to sort through the details to identify common threads and patterns and then to share these patterns with your client.

- Deductive reasoners tend to have a rigid framework from which they view reality. They tend to disregard important information that does not fit into their framework (e.g., "I can't have this condition, I'm a healthy person!"), and only take in information that supports their view. You will need to be patient with clients who have rigid frameworks, helping them to see that reality is not so clear-cut. Deductive reasoners may be particularly frustrated with the uncertainty that is inherent in genetic counseling (e.g., "What do you mean you can't tell me how severely affected my child will be?").

- Synthesizers can take in information that both confirms and challenges their frameworks. Synthesizers can absorb a great deal of information rather easily and use it to make decisions. They tend to spend most of their information processing time in their heads to the neglect of their feelings. You will need to encourage synthesizers to discuss their emotions.

- Confused reasoners are not less intelligent than other clients, but they have never learned how to process information. They may experience intellectual confusion because they cannot differentiate between important and trivial information. Confused reasoners may spend a great deal of time on a small point, while missing the larger issue, for example, the client who wonders if her unborn child who has Down syndrome will look like her, while missing the point that the child will be cognitively impaired; or the client who wants to know all of the statistics associated with any report on his disorder and yet is unable to discuss how it has and will affect his life.

Emotional Styles

- Spontaneous style: These types of clients are active communicators, who respond easily and expressively. Often they have a good sense of humor that they will use even when they are sad (e.g., joking while crying). They will tend to use both humor and denial as defense mechanisms. You must attend carefully to these clients because they may portray everything as being fine when it is not. Also, you can help them "own" their reactions by gently encouraging them to express their true feelings.

- Nonexpressive style: These types of clients are very articulate, but in a highly intellectual way. Although they feel their emotions, they deny that their feelings have any importance. They may even express some annoyance or disregard for individuals who do show their emotions. They appear to be confident and in control. You should try to moderate the intensity of your own emotional expression, because these types of clients tend to shut down even further when other people are very emotive.

- Reserved style: These clients express their feelings to a limited extent, but do not allow their full expression. You can usually draw out these clients somewhat with primary and advanced empathy reflections of feelings and with questions (e.g., "Please tell me more about what it's like for you to be so sad"). Then, after your client discloses, you can ask what the client will do to cope with her or his feelings.

- Explosive style: These clients express everything that they are feeling, sometimes in overly intense ways. They can be demanding, histrionic (dramatic), and may lack good interpersonal boundaries. You need to set clear limits (e.g., regarding session length, physical contact, what you will and will not provide, etc.). It is also important that you remain calm if they have outbursts, such as crying hysterically.

Spiritual Dimension

The spiritual dimension pertains to client beliefs in a higher power and includes personal values, practices, personal philosophy of life, and connection to religious institutions. Note that religion is only one part of this multifaceted dimension.

Clients' spiritual beliefs and practices affect their interpretation of genetic information and subsequent decisions. For example, Greeson and colleagues (2001) interviewed Somali immigrants who were Muslim and found that their religious beliefs profoundly influenced their perceptions of the causes and consequences of disability. The authors concluded that these religious views would have a significant impact on the utility of genetic services.

You should consider the possibility that spiritual issues are relevant to some extent for most of your clients, even those clients who do not mention them. You should directly inquire about their importance. For example, you could ask:

- What religious/spiritual beliefs do you hold, and how do they relate to your concern?

- What values do you hold?

- Would you want your beliefs to be part of your decision-making process?

The counselor need not be of the same faith or embrace the same belief system in order to be of help to the client. It is not necessary for the counselor and client to share the same religious faith, any more than they need to share the same sex, race, or background. What is important is that the counselor have an open demeanor, accept differences, and be able to enter the client's inner world. [Cheston, 1991, p. 127]

Closing Comments

In the other chapters in this manual, you read about and practiced several basic helping skills. Part of the art of being an effective genetic counselor involves timing and choosing, that is, knowing when and how to intervene. Different clients require different choices on your part. This chapter described a few ways in which clients differ stylistically, discussed some of the client emotional issues that you will encounter, and discussed how clients may resist and defend when they feel threatened. Clients differ in many other ways, more than we could adequately cover in one chapter or even one book. As you gain supervised experience, you will increase your sensitivity to client differences and learn to more effectively individualize your counseling approaches.

Class Activities

Activity 1: Resistance Role-Plays

Students work in triads, each taking a turn being the genetic counselor, client, and observer. They should engage in 10-minute role-plays in which the client presents a very specific issue (e.g., advanced maternal age and is trying to decide whether or not to have an amniocentesis. There are no other known risk factors). The client and counselor should act out this issue using one of the following resistance role-plays:

ROLE-PLAY 1

When the genetic counselor provides some information about the client's options, the client should respond to all counselor suggestions with "Yes, but. . ." In other words, the client should refute, counter, and find fault with all counselor options.

ROLE-PLAY 2

The client should be silent, giving one or two word answers, refusing to answer, and trailing off in her responses. The client is making it evident that nothing is going to happen until the counselor addresses her/his resistance (which is due to the fact that the client was told by her doctor that she could have an amniocentesis only if she first went to genetic counseling; also, she does not want to discuss her personal history with a stranger).

ROLE-PLAY 3

The client should get angry in response to the genetic counselor's first question.

PROCESS

Discuss with the counselor how it felt to be resisted in this way. What did the counselor do to respond to the client's resistance? Why might clients use these types of resistance?

Estimated time: 60 to 75 minutes.

INSTRUCTOR NOTE

- To make this activity more challenging, the student playing the client could select the role-play without telling the individual playing the counselor which role s/he selected.

Activity 2: Magician Scarves Discussion

Hopkins (1997) uses the "magician's scarves" as a metaphor for grief. The magician pulls one scarf after another from his or her sleeve in what seems to be an endless process. It seems impossible that the magician could have so many scarves. But eventually the magician stops and takes a bow.

Grief is like those scarves. Once we open ourselves up to the experience of grieving over one loss, we begin to feel the pain of an earlier loss. Sometimes it feels as if the losses will never stop emerging. But, like the magician's scarves, there is a finite number. Some are large and brilliantly colored, others are smaller and more neutral in tone, but each can be experienced and resolved. [Hopkins, 1997, p. 24]

Using the magician's scarf metaphor, students do the following: With different colored magic markers, draw scarves on a large piece of paper. They can begin anywhere and draw the first scarf to represent the first loss that comes to their mind. It may be large or small, neutral or vividly colored. As they draw, attach the scarves to each other.

After they finish drawing, they discuss their drawings with a partner of their choice. Explain what each scarf represents. (Students should discuss only those scarves representing losses that they are comfortable discussing.)

Estimated time: 50 to 60 minutes.

- You may need to provide some examples of losses in order to get students started, such as moving to a new city, losing your best friend, losing in a sporting competition, loss of important mentors or family members, etc.
- A variation of this exercise is to give students a brief client scenario (e.g., prenatal diagnosis of translocation with subsequent pregnancy termination). The students could draw one picture representing the mother's losses and another picture representing the father's losses. Then they would discuss their pictures with a partner or with a small group.

Activity 3: Defense Mechanisms Small or Large Group

Using the list of defense mechanisms in Table 9.1, students generate additional examples of client statements or behaviors to illustrate each one. They can first work in Think-Pair-Share dyads to generate ideas.

Estimated time: 20 to 30 minutes.

INSTRUCTOR NOTES

- Following the generation of examples, the instructor could lead a discussion of which defenses each student finds particularly challenging to work with in genetic counseling.
- Students could generate counselor responses to address each client statement or behavior.

Activity 4: Role-Playing with a Grieving Client

Students work in triads, each taking a turn as counselor, client, and observer. Using the following client roles, they should engage in 15-minute role-plays in which they discuss the client's affect.

CLIENT ROLE 1

A woman just found out from her routine ultrasound that the fetus died.

CLIENT ROLE 2

A 50-year-old woman was told by the neurologist that she has symptoms of Huntington's disease.

CLIENT ROLE 3

A mother of a 6-year-old boy found out last week that testing showed her son to be affected with Duchenne muscular dystrophy.

PROCESS

Students discuss in the large group: What are you learning about client grief and how you respond to it? What is difficult about it? What is the genetic counselor's role in addressing client grief? How did you feel responding to strong client emotion?

Estimated time: 75 to 90 minutes.

Written Exercises

Exercise 1: Grief and Loss

Describe a situation in your life where you experienced a significant loss. Do the aspects of grief described in this chapter accurately represent the process you went through to cope with your loss? How do they fit? Not fit? What sorts of things do you recall people saying to you at the time of your loss that were especially helpful? Unhelpful? Recommended length: one to two word-processed pages, double spaced.

Exercise 2: Anger

Part I: Describe the meaning of anger in your family of origin. For example, was it a socially acceptable emotion? What did it mean when someone became angry? How was anger expressed? How did others react to it? How does the meaning of anger within your family compare to the meaning within your culture? How do you currently react to anger? How do you currently express anger?

Part II: If you have had a genetic counseling client become angry with you during a session, describe what happened, how you felt, and what you did. In retrospect, do you wish you had done something differently? If so, what? (If you have not actually had this experience, then make up a scenario and use it to respond to these questions.)

Exercise 3*: Anger

Read the following scenario and respond to the questions: pretend that you are the clinical supervisor of an advanced genetic counseling student. During a supervision session, your student says, "You asked what I think I do well as a genetic counselor. Well, the clients I see in genetic counseling never get angry with me, or show much anger about anything."

QUESTIONS

What is your reaction to your supervisee's statement? What would you say

---

*Adapted from Cavanagh, 1990.

to the student? (Write out what you say as if you were talking directly to her/him.)

Exercise 4: Defense Mechanisms

Using the defense mechanisms in Table 9.1, identify one or two defenses that you are most likely to use and discuss how they might affect your work as a genetic counselor.

(Hint: Specific examples of how they might play out during a session and how they would affect the client would be helpful.)

Exercise 5: Intellectual and Emotional Styles

Part I: Using the intellectual styles described in this chapter, identify your intellectual style and discuss the advantages and disadvantages of your style for genetic counseling. Do you think that your style might be more effective for some clients and less effective for others?

Part II: Using the emotional styles described in this chapter, identify your emotional style and discuss the advantages and disadvantages of your style for genetic counseling. Do you think that your style might be more effective for some clients and less effective for others?

(Hint: Think about how your intellectual and emotional styles might complement or clash with the client's intellectual and emotional styles.)

Part III: Which client intellectual and emotional styles will be most difficult for you? What makes them difficult?

Annotated Bibliography

Clark, A. J. (1991). The identification and modification of defense mechanisms in counseling. Journal of Counseling and Development, 69, 231-236.
[Describes different mental health client defense mechanisms and offers some suggestions about responding to them.]
Djurdjinovic, L. (1998). Psychosocial counseling. In: Baker, D. L., Schuette, J. L., Uhlmann, W. R., eds. A guide to genetic counseling. New York: John Wiley and Sons.
[Discusses typical genetic counseling client reactions to bad news and challenging situations, including anger, grief, guilt and shame, and denial.]
Hultman, K. E. (1976). Values as defenses. Personnel and Guidance Journal, 54, 269-271.
[Identifies what he terms *defensive values* for mental health clients; describes criteria for determining when values actually are defenses; and discusses why they are problematic with respect to goal setting and decision making.]
Kessler, S., Kessler, H., Ward, P. (1994). Psychological aspects of genetic counseling. III. Management of guilt and shame. American Journal of Medical Genetics, 17, 673-697.
[Discusses client guilt and shame and offers practical suggestions for genetic counselor interventions.]
Murray, R. F., Jr. (1976). Psychosocial aspects of genetic counseling. Social Work in Health Care, 2, 13-23.

[Provides clear descriptions of the types of defense mechanisms used by clients who have genetic conditions or whose children have a genetic condition. Also discusses some client affect. A practical article, although some of the attitudes about the role of mothers are dated.]

Ormerod, J. J., and Huebner, S. (1988). Crisis intervention: facilitating parental acceptance of a child's handicap. Psychology in the Schools, 25, 422-428.

[Written from the perspective of the school psychologist, this article provides a clear description of parental reaction to the diagnosis of a child with a handicap and offers specific crisis interventions that service providers can use to address parental reactions. Many of these suggestions are appropriate for the genetic counseling setting.]

Rich, D. E. (1999). When your client's baby dies. Journal of Couples Therapy, 8, 49-60.

[Describes the grief reactions for couples who experience perinatal loss, suggests therapeutic interventions, and discusses counselor countertransference.]

CHAPTER 10

Counselor Self-Reference:
Advice Giving, Self-Disclosure,
and Self-Involving Responses

Learning Objectives

1. Define advice giving, self-disclosure, and self-involving skills.
2. Differentiate self-disclosure from self-involving responses.
3. Determine guidelines for effective advice giving, self-disclosure, and self-involving responses.
4. Identify examples of counselor-client themes appropriate for self-involving responses.
5. Develop advice giving, self-disclosure, and self-involving skills through practice and feedback.

In Chapter 2 we discussed how Rogers' person-centered model of helping forms the theoretical basis of genetic counseling. In Chapter 11, we reiterate the role of nondirectiveness as part of the ethical guidelines for genetic counseling but also describe situations in which genetic counselors may take a more directive stance. In this chapter we discuss three genetic counselor skills that involve more directive behaviors: advice giving, self-disclosure, and self-involving responses. Advice typically refers to statements of the genetic counselor's professional opinions, while self-disclosure and self-involving responses are revelations of the genetic counselor's personal experiences and reactions. These are infrequently occurring but potentially very powerful genetic counselor interventions. We begin with a discussion of advice.

Advice Giving

Definition of Advice

Advice is a type of response in which the counselor attempts to directly influence clients by offering suggestions, recommendations, or opinions about what they should do. As opposed to information giving, which involves the communication of knowledge (see Chapter 7), advice involves a recommendation about a particular course of action. Furthermore, "With advice, the giver has a stake in it, and the receiver is expected to act on it [whereas] we give information generously, without strings, and the client is free to use it or not, to use it as he sees fit, or to file it away for another time. Advice belongs to the person who gives it; information belongs to the person who gets it" (Fine and Glasser, 1996, p. 66). Kessler (1992a) argues, "In rendering advice or informing a counselee as to what their personal choice might be in a given situation, the counselor overtly attempts to shape and influence the counselee's behavior" (p. 10). Advice is given in order to offer recommendations, to help advice seekers sort through the options that they have already decided on, or to help them implement their decisions successfully (DeCapua and Findlay Dunham, 1993).

Appropriateness of Advice Giving in Genetic Counseling

Advice giving is a controversial skill for genetic counselors. We believe that the genetic counseling field is caught between the practices of the two professions with which it shares a number of similarities: medicine and mental health counseling. In medicine, physicians and other health care providers may routinely offer opinions or recommend a certain course of treatment, while in mental health counseling, practitioners typically avoid giving advice.

To further complicate matters, clients may expect you to provide advice because they often regard genetic counselors as experts on genetic and medical topics, and will therefore assume "that the counselor will be an advice giver, will know what is 'best' to do in a particular circumstance, and will make a recommendation that must be followed" (Baker, 1998, p. 70).

Kessler (1997b) cautions that the provision of advice "is often a vote of no confidence in the client's own ability to sort things out for themselves and arrive at their own conclusions. It needs to be remembered that most of the people seen for genetic counseling are experienced decision makers; they have already made multiple decisions in the course of their lives without our help" (p. 383).

A number of authors in the mental health field similarly caution against advice. For example, in crisis intervention counseling, there is a danger that advice may impose counselor values and preferences on clients (Gravely Moss, 1985). Advice givers often begin by saying, "If I were you..." However, one problem with this statement is that "counselors are not clients and cannot know what choices are best for them. Counselors do not know all the past experiences that have influenced clients' present crisis states, their thinking or feelings, and therefore are not qualified to make suggestions about many problems and concerns" (Gravely Moss, 1985, p. 12). Gravely Moss suggests that instead of offering advice, counselors explore with clients the positive and negative aspects of the different options that they are considering, because the goal of crisis intervention is to "help clients identify decisions they feel comfortable making to solve problems" (p. 13). This goal seems quite appropriate for genetic counseling, which shares some similarities with crisis intervention counseling such as time-limited counselor/client contact, clients who need to make a decision(s), and clients who may be in a highly emotional state.

Hill and O'Brien (1999) warn that beginning mental health counselors are prone to giving too much advice or giving it too soon: "It is critical to realize that the need to provide answers often originates in the helper's insecurity and desire to help, which are normal feelings at the start of learning helping skills" (p. 74).

Consequences of Advice Giving

Some authors point out that advice may have both positive and negative consequences within interpersonal relationships: "Advice may be seen as helpful and caring or as butting in; advice may be experienced as honest or supportive; and seeking and taking advice may enact respect and gratitude, yet recipients reserve the right to make their own decisions" (Goldsmith and Fitch, 1997, p. 454).

POSSIBLE POSITIVE OUTCOMES OF ADVICE

Based on family therapy literature, when you give advice successfully, you may offer suggestions that clients perceive as helpful, you may present a new idea that they had not considered before, and your advice may give permission to take an action that the clients wanted to take anyway (Silver, 1991). Advice may provide informational support and directive guidance, and it may demonstrate caring and give the impression that a problem is manageable (Goldsmith and Fitch, 1997).

POSSIBLE NEGATIVE OUTCOMES OF ADVICE

Advice may also have negative effects. Clients may feel criticized because advice indicates that they should be doing something differently; they may feel constrained to consider only the options that you raise; they may feel pressured to follow your advice; and they may become oppositional (i.e., resisting everything you say for the remainder of the session) (Silver, 1991). Additionally, advice may put the recipient in the position of appearing dependent and less capable if s/he takes the advice, and of appearing disrespectful and ungrateful is s/he does not heed the advice (Goldsmith and Fitch, 1997). Another risk of giving advice is that your client may blame you if the advice doesn't work.

TYPES OF ADVICE

Genetic counseling often involves certain types of advice, such as (1) standards of care (e.g., "If you want to know if you have the gene for Huntington's, then this is the testing that you should have done"; or if a client needs to see a specialist, but doesn't know how to go about finding one, it is appropriate to give your opinion on how to proceed); (2) medical recommendations (e.g., "I'd recommend that you have regular mammograms because of your family history of breast cancer"); (3) the genetic counseling process (e.g., "To help you come to a decision that works for you, I suggest that you and I talk through the different options"); and (4) client behavior (e.g., "I think that you should take a couple of days to think this over"; or "You might benefit from talking to some of the other parents in the local muscular dystrophy support group"). However, genetic counseling rarely involves telling a client what decision to make; for example, "I think that you should have an abortion."

Suggestions for Giving Advice

Make it clear that advice is not routinely part of genetic counseling. At the beginning of the session, when you explain the process of genetic counseling, state that you tend to refrain from telling clients what to do because you prefer to help them come to a decision for themselves. This helps to limit any subsequent advice giving or requests for advice from your clients.

Give advice later in the session. You should offer advice only after you have demonstrated that you have some expertise in the topic, after you have established rapport and have shown that you care, and only if appropriate to the situation. Also, you should wait until you have listened fully to the client's situation and have demonstrated accurate empathy (Goldsmith and Fitch, 1997).

Offer advice tentatively. You should avoid getting into arguments with clients. Back down if a client resists, and do not try to "argue the client into backing down and accepting what you say" (Martin, 2000, p. 63). Arguing is almost never effective for influencing clients to follow a suggested course of action. Instead, they will shut down, pretend to agree, leave prematurely, etc.

Mention decisions made by other clients. Sometimes it is helpful to briefly and anonymously describe what other clients in similar situations considered to be their options.

Check out the impact of your advice. Ask clients to discuss what they think and feel about your recommendations. This also helps to ensure that they accurately understood your advice.

Try questions instead of advice. Use questions that get at what the client considers to be the pros and cons of different options instead of suggesting what the client should do (Silver, 1991). For example, you might say, "If you chose to have the baby, what would happen?"; or "Which decision do you think would make you feel more comfortable or at peace?" You will assist clients in thinking through their decision-making process to arrive at the outcome that is best for them rather than directing them to do what you think is best.

Emphasize the decision-making process rather than the outcome. "You've asked me if I think you should have an abortion. From what you've said, I'm thinking that these are some of the reasons you might make that decision, and these are some of the reasons that you might not…"Such a response provides what Kessler (1997b) refers to as a framework from which clients can view things more clearly.

Be culturally sensitive. Remember that clients differ in their desire for advice. Sometimes these differences are due to cultural background. For example, Silver (1991) cites research from the mental health field indicating that clients from lower socioeconomic classes expect to receive advice from their therapists and are more satisfied when they receive it. Goldsmith and Fitch (1997) cite research suggesting that individuals who are members of the following cultural groups may desire advice: Germans, who value frequent, direct advice, especially among friends; Colombians, who value high involvement as demonstrated by frequent and explicit acts directing others' behavior; and Israeli Sabras, who value advice about individual behavior aimed at directing this behavior toward the good of the community. Some clients from Asian cultures may similarly expect guidance and direction (Ishiyama, 1995).

You may need to explain that rather than providing advice, you will provide as much pertinent information as you can and possibly suggest some ways that clients could go about using this information to arrive at a decision that is best for them. Such a statement can help to reduce misperceptions for any client who expects you to be highly directive.

Diller (1986) suggests the following steps for physicians to give advice successfully, which we adapted for genetic counseling:

- Determine the client's view of the problem (e.g., "What are you worried about? What do you know about the condition? Do you know anyone who has had it? What have you thought to do about it? What do you think would work? Would not work? What have others told you about what to do? What do you think about their advice?").

- Use primary empathy to reflect your understanding of your client's view of the problem (e.g., "It seems like you feel very torn about what to do").

- Locate any hidden, explosive issues before giving specific advice (e.g., "What options do you think you have?" This question could prompt a client to talk about how abortion is morally offensive to her and that it is something she would never consider).

- Acknowledge the client's ambivalence (e.g., point out how all of the options have risks and uncertainties associated with them and therefore it is hard to make a decision).

- Give advice in language that is consistent with the client's view of the problem (e.g., "You've said several times that you feel like you are 'being smothered' with the weight of this decision. What would you think about taking a couple of days to catch your breath? Could you put down the weight for a little while before you make your final decision?").

Advice-Giving Pitfalls

There are a number of factors, some pertaining to the genetic counselor, some to the client, and some to the situation, that can lead to ineffective advice giving. It is important to be aware of the following:

Giving Advice to Satisfy Your Own Needs

As Kessler (1992a) eloquently states, "Some [genetic] counselors have the fantasy and wish that if they could only exert their personal power of persuasion, others will begin to see the world the way they do.... Perhaps genetic counselors need to learn what others engaged in personal counseling and psychotherapy have had to resign themselves to and that is we are not very powerful when it comes to changing the behavior of others" (p. 16). Advice givers often feel powerful, helpful, and competent when they give advice (Silver, 1991). Since these outcomes feel quite good, they may tempt you to give more advice than you should.

Giving Advice Based on Faulty Assumptions

Silver (1991) suggests that mental health counselors who give advice may hold one or more of the following beliefs:

- Professionals know what is best.
- Clients do not know what is best.
- Professionals should take responsibility and make decisions for clients.
- Clients can't take responsibility for making their own decisions.
- There is one best view and solution, and the professional knows what these are.
- Clients want advice.
- Clients benefit from advice.
- An objective third party is in the best position to give advice.

Mistakenly Thinking that Clients Are Seeking Your Advice

Clients often have already made up their minds before asking your opinion. They are not actually asking for advice; rather, they are seeking support for their decision (Goldsmith and Fitch, 1997). However, they seldom come out and directly ask for this support. Instead, they will disguise it in the form of a request for advice. For these individuals, giving advice, especially advice that is discrepant with what they have decided, could be perceived as offensive, and they certainly would not consider it (Kessler, 1992a). Additionally, what sounds like a request for advice may actually be a request for information that would allow the client to make her or his own decision (Kessler, 1997b): "Commonly this information concerns a way to think about a problem rather than a solution" (p. 383).

Putting Your Expertise on the Line Prematurely

As we said earlier, you should refrain from offering advice until you understand the situation more fully and have enough experience to be able to offer an expert perspective. As Pedersen and Ivey (1993) point out, "In many cultures counseling is frequently done by persons in a teaching role, and the efficacy of a counselor is judged by the truth and wisdom of his or her teachings" (p. 189).

Thinking that Clients Will Listen to Your Advice

Clients usually will not take your advice. They might act as if they agree, but privately they have discounted your suggestions. Or they might say, "Yes, but..." and go on to explain why your advice won't work. Fine and Glasser (1996) point out, "What a person tells himself is more valuable to him than anything you might tell him, even if what you tell him is better" (p. 66).

Not Realizing that You May Appear to Be Taking Sides

Advice will probably lead you to be identified with the family member(s) or friend(s) who has already made a similar suggestion (Silver, 1991).

Believing that You Know Better Than Your Clients What They Need to Do

While you may be an expert on genetic counseling, you are not an expert on your client (Cavanagh, 1990). Consider for a moment how much you think a person would know about you after spending 1 hour together. Furthermore, "no amount of empathy can replace the fact that the counselee has to make and live with their decision" (Kessler, 1992a, p. 14). As you gain experience you will begin to get a feel for typical or normative types of client reactions and decisions. You must remember, however, that a typical reaction may not fit for the particular client sitting in front of you.

Forgetting that Clients Ultimately Are Responsible for Making Their Own Decisions

When you provide advice, you risk shifting the responsibility for the outcome to you, especially for clients who appear to be desperate for advice. Instead, try indicating that you understand their need for advice and are willing to help them figure out what's best for themselves (Martin, 2000). For example: "I know that you feel like you can't make this decision alone. Why don't we try together to figure out the best way to proceed?"

Thinking that Your Behavior Is Advice-Free

Clients who want advice will believe that you've given it even when you thought you were being nondirective (Silver, 1991). It is important to watch for clues that clients are trying to pull a recommendation from you (e.g., client says, "You probably think this is a bad idea..."; or "I suppose you think I'm making a mistake..." Think carefully about how you want to respond to such statements).

Examples of Genetic Counselor Advice

Here are several examples of the types of advice that experienced genetic counselors have given to actual clients:

- Some clients say that they wish they had taken more time to make the decision to terminate a pregnancy. So when counseling a client(s) who is at the decision-making stage, the counselor says, "I know it is natural to want to make a decision quickly and move on, but I would encourage you to take a few days."

- "Consider the issues very carefully. Take as much time as you need."

- A couple was in conflict with each other over whether or not to have an amniocentesis. After discussing the issue in the genetic counseling session, the counselor said, "I would encourage you to continue this conversation at home, look at both of your opinions about this, and take some time to make the decision."

- "Why don't you wait until your husband gets here and then maybe you will feel more comfortable about making a decision."

- Fragile X female carriers often feel alone, guilty, and burdened. Recognizing this, the counselor said: "I would encourage you to include your husband in these decisions about having more children and about how you would handle the child care."

- "I think it's important that you know about all of the options available to you."

- "I think it would be a good idea to find out for sure whether or not your uncle has trisomy 21 or the translocation form of Down syndrome before we decide whether or not to test your chromosomes or have an amniocentesis."

- "It sounds to me like you would prefer to have the ultrasound before we discuss the option of amniocentesis any further. I think that sounds like a good idea."

- "In making a decision about carrier testing, it is important to decide whether the information will be useful to you. If you know your child were at risk for cystic fibrosis (CF), would you do anything or plan anything differently?"

- "One thing to think about when deciding about amnio is your own level of concern about the conditions it can detect, and compare that to your level of concern about the risk of the procedure."

- "One thing you might consider is having an autopsy done to help you assess the risks of this happening again."

- The client stated that she wanted an amnio because her doctor told her to have the test. The counselor said, "Perhaps you should focus on what is best for you and your family and not on what the doctor is recommending."
- "Many of my clients have found it very helpful to talk to other parents of children who have ___."
- "I recommend that you subscribe to the ___ newsletter to stay current about new developments, changes in health care management, etc."
- "If you don't feel like you can give up alcohol on you own, then we should get you some help. It is very important to avoid alcohol during pregnancy."

Self-Disclosure and Self-Involving Responses

Two types of responses involve personal revelations by genetic counselors: self-disclosure and self-involving responses. These responses vary in the extent to which they are directive (e.g., high in directiveness, telling a client what you would do in his or her situation, versus low in directiveness, saying that you have had experience working with other clients who have CF). They also vary in their intimacy level.

Definitions of Self-Disclosure and Self-Involving Responses

Self-disclosure is the counselor's communication to the client of information about her- or himself. It includes a range of experiences such as beliefs, attitudes, perceptions, judgments, desires, and actions, as well as feelings about people and/or situations other than the client.

Self-involving responses are direct communications of the counselor's feelings about and reactions to the client in the here-and-now situation.

Self-Disclosure

Counselor self-disclosure is a controversial behavior. Within genetic counseling, some authors (e.g., Kessler, 1992b) suggest caution in its use, arguing that self-disclosure is highly directive (e.g., telling a client what you would do if your were in her or his situation), and since there is little research evidence concerning the impact of this type of directiveness, genetic counselors should be cautious with respect to its use. Some authors in the mental health field (e.g., Simone et al, 1998), caution that self-disclosure may reflect the therapist's unconscious needs (e.g., for intimacy), and it may blur relationship boundaries (you are a counselor, not a friend).

It is our view that a certain amount of self-disclosure is always present, and therefore the issue is not whether to disclose, but rather how much and what ways to directly share yourself in the genetic counseling relationship.

When we say that some self-disclosure is always present, we mean that it is impossible not to reveal some information about yourself. Every behavior that you engage in with a client (your facial expressions, posture, voice tone, etc.)

reveals something about you. Furthermore, clients may *read into* your behaviors. They may determine (not always accurately) whether you approve or disapprove of their behavior, whether you would or would not make the same decision, etc.

Your personal characteristics also communicate information that clients will actively interpret:

- Gender (e.g., a prenatal client may believe that you can understand her situation if you also are a woman, and may question your ability to understand if you are a man).
- Age (e.g., you appear to be too young to know what it's like to have experienced years of infertility.)
- Race/ethnicity (e.g., the client may feel some distance if you appear to be of a different ethnic group).
- Marital status (e.g., if you wear a wedding ring, you may be perceived as someone who understands the conflict a couple is experiencing).
- Physical appearance (e.g., if your physical characteristics suggest that you have a genetic condition, this might make some clients reluctant to discuss termination of an affected pregnancy).
- Office decorations (e.g., pictures of your children may lead some clients to perceive you as understanding about pregnancy, but perhaps less about pregnancy loss).

This type of *indirect* disclosure is not something that you can or necessarily should try to control or manipulate. But it is important to think about what your characteristics and actions may convey to clients.

Another category of self-disclosure is *direct* disclosure. This type of self-disclosure is the focus of this chapter. Direct disclosures are intentional communications about yourself. Sometimes you will deliberately choose to reveal information about yourself, and sometimes your disclosures will be in response to client questions (e.g., "Have you ever worked with clients who have a genetic condition like mine?").

Direct disclosures vary along a continuum from low, to moderate, to high intimacy. In Table 10.1 we describe and illustrate different levels of disclosure intimacy.

Functions of Self-Disclosure

McCarthy and Oakes (1998) identify several reasons that mental health counselors and other human service professionals might disclose. Several of these are relevant for genetic counseling:

Enhances Social Influence

Counselors appear to be more human when they self-disclose; they may be regarded as more receptive, warm, capable, and trustworthy (McCarthy 1979, 1982). Examples include:

TABLE 10.1. Self-disclosure levels of intimacy

Self-disclosure	Example
Low intimacy	
Situationally required	I am a second year master's student.
Beliefs, opinions	I believe it's important to take your time making a decision like this.
Facts about yourself	I am married.
Personal feelings and perceptions not generated by the present relationship	I feel frustrated when test results are inconclusive.
Expressions of feelings experienced in the past toward the client	I was concerned after talking with you last week.
High intimacy	

- "You look uncomfortable/anxious. I feel that way, too, when I go to the clinic."
- "I have a strong family history of heart disease, too. I've been working on changing my diet and exercise, but it sure isn't easy!"
- Clients often ask about the pictures on a counselor's desk. "Do you have children?" or "Are these your children?" The counselor acknowledges it, and reveals a small amount of information—their ages, for example.
- Client: "I'm worried about what these test results will do to my children." Counselor: "I have children of my own and can understand why you feel that your decision also affects them."
- "I have been working in this clinic more than 10 years."

Builds the Relationship

Self-disclosure can communicate that you are interested and concerned. It suggests that you are genuine, and it implies that you trust the client with the information that you reveal about yourself. Examples include:

- "When I was pregnant, I was anxious/worried all the time."
- "So you are from Houston. I grew up there. I really miss___."
- Identifying with your client as a parent: "I think it's natural for us as parents to struggle when deciding what's best for our families."
- Client: "This pregnancy is so important to me. We spent 5 years getting pregnant and had to do an in vitro fertilization. Counselor: "Infertility can color your perspective and make any type of risk to the pregnancy seem unacceptable. I felt that way, too, when I was deciding about amnio."

REINFORCES CLIENT SELF-DISCLOSURE

There is a give and take in that both the client and you are risking being open.

- Cl: "It's hard to explain what's going through my mind right now."
  Co: "Sometimes I have a hard time explaining myself when my feelings are so mixed. I wonder if that is how it is for you?"

REASSURANCE THAT THE CLIENT IS NOT ALONE

This type of disclosure tends to normalize or validate client feelings and decreases client anxiety and isolation.

- Cl: "I feel so mixed up right now; I should be able to make this decision."
  Co: "I'd have a difficult time, too, if I were in your situation."

PROVIDES REALITY TESTING

Disclosure can point out that decisions are not always clear-cut.

- Cl: "I can't believe I'm waffling on this! I mean, I told you before the amnio that I would never consider terminating a pregnancy! That just seemed so wrong. But now I don't know."
  Co: "It's been my experience that many people in a similar situation feel as conflicted as you do."

GENERATES NEW PERSPECTIVES

You may be able to assist clients in decision making by providing a different viewpoint or strategy.

- Cl: What would you do if you were me?
  Co: I'm not sure, but when I've had a big decision to make, I think things over for a while and talk with my family. I try not to rush into a decision before I feel ready.

ELICITS STRONG FEELINGS

Clients may become sad, angry, and/or frightened during the session. You may need to give them permission to express these feelings. One way to do this is with self-disclosure. Examples include:

- Co: "I know that if I were in your situation, I'd be feeling very sad right now. Is that how you are feeling?"
- Co: "It looks like you want to cry and that you're holding yourself back. Maybe because you're embarrassed? Sometimes I feel uncomfortable crying, but I know it helps me to let it out."

GUIDELINES FOR USING SELF-DISCLOSURE

McCarthy and Oakes (1998) offer the following guidelines for using self-disclosure:

Examine your reasons for disclosing. Self-disclosure should help clients accomplish their goals for genetic counseling. Before disclosing, ask yourself: Will my disclosure help the client open up, see a different perspective, or move toward a decision? Consider if your disclosure is intended "to foster the client's learning or is it a subtle solicitation for appreciation, admiration, or sympathy for yourself?" (Fontaine and Hammond, 1994, p. 224).

Be intentional. Keep your disclosure brief and focused. A useful technique is to make a disclosure and then immediately follow it with a question, "Is that how it is for you?" or "What do you think (or how do you feel) about what I just said?" This brings the focus back to your client and helps the client use the information for his or her own situation.

Choose an appropriate intimacy level. Since disclosures range on a continuum from demographics to highly personal experiences, you must be sensitive enough to choose a level that will not overwhelm or alienate your client. One way that you might gauge the intimacy of a disclosure is to ask yourself how many other individuals you have told this information to in the past.

Choose an appropriate time. If you disclose too soon, you allow clients to avoid the painful work of making decisions for themselves. You risk having them grab onto the solutions that you or your previous clients have generated. They do not work it through, and they risk selecting an option that is not the best for themselves. Also, be careful about disclosing before you have begun to understand your client's frame of reference, and, in general, you should save more intimate disclosures for later in the session.

Be conservative. Too frequent, too intense, and/or too lengthy disclosures can shift the focus to you, can burden your client with your problems, and can be distracting.

Keep more personal disclosures nonimmediate. Some research suggests that disclosure of a current crisis leads some clients to perceive mental health counselors as less adjusted and less helpful (Knox et al, 1997). In general, it is less risky to disclose information from your past experiences rather than from your current experiences. For example, clients might be distressed to learn that you currently are undergoing testing for Huntington disease, but might benefit from learning that your underwent testing 5 years ago.

Know your own reactions. You should anticipate how you will feel as you disclose certain information and generally avoid topics that would cause you to be highly emotional. Before you begin seeing clients, think about what information you might reveal and what information you would never reveal. You should avoid revealing information that would make you highly distressed, anxious, angry, etc. On a related note, it is important to remember that although you are required to maintain client confidentiality, your clients are not under the same imperative. Clients can reveal anything that they care to, and with anyone. Therefore, in thinking about topics that you might disclose, you should consider how you would feel if clients passed this information along to others. Furthermore, clients vary in how sensitive and nonjudgmental they are about your situation.

Anticipate client reactions. Before disclosing, ask yourself how you think this information will affect this client. Only disclose if you believe that it will be helpful to your client in achieving her or his genetic counseling goals. When disclosure is requested by a client, first try to determine the motivation behind the request. Does your client really want the information? Is your client hoping for a particular answer? Is your client actually saying that s/he doesn't trust her or his own judgment? Depending on the motivation, one of the following responses might be appropriate:

- "I'd be happy to tell you what my experience has been, but first, I'm wondering, What prompted you to ask?"
- "What are you hoping that I will say?"
- "I wonder if you are hoping that I will be able to give you the right answer. What I would do in your situation may not be appropriate for you. Let's try together to figure out which option is best for you."

In Chapter 7 we described dependent decision makers who rely on others to make their decisions for them. Be watchful for this motivation if a client asks you what you would do. If you decide not to disclose, then you might try the strategy recommended by Kessler (1992b). He suggests that genetic counselors avoid answering questions about what to do by redirecting attention to process issues (e.g., a young genetic counselor responded to questions about whether or not she had children by stating that perhaps the client was wondering if she would be able to understand). If you do choose to disclose, consider not only saying what you would do if asked, but also describing your decision making process in order to help clients understand (and perhaps use) this process (Bell, 1990).

Consider the client's characteristics. Some multicultural experts advocate self-disclosure with clients from cultures where who you are is more important than what you do (Vontress, 1988). So, for example, clients from Puerto Rico might ask direct questions about your marital status and whether you have children. You should consider answering some direct questions in order to build a trusting relationship. Conversely, self-disclosure may be contrary to the basic values of some cultural groups. For instance, there are often lower self-disclosure rates in Eastern cultures as compared to Western cultures. Try to become familiar with the disclosure norms of different cultures.

Examples of Genetic Counselor Self-Disclosure

Here are several examples of the types of self-disclosures that experienced genetic counselors have made to their clients:

- "Oh, my. You have that spitting condition. My sister had it, too. She was miserable. How are you coping?"
- "I think that, too. Sometimes all the technology we have today just makes things more difficult. It sure would have been simpler to be pregnant 100 years ago."
- "I think that I would also have a difficult time deciding what to do."
- "I am also a very fact-oriented/information-seeking person, so I can understand your need to know everything possible before making a decision."
- "I don't have children yet, so I don't know first hand what it's like to have to make a pregnancy decision. But I have walked through similar situations with my clients in the past."
- "I know how hard it can be to deal with insurance companies, too. I've run into roadblocks with them myself. I can help you make some progress in getting this procedure/test covered."
- "My mother had breast cancer, and I remember how difficult it was to live with uncertainty. How is it for you?"
- "I am Catholic as well, and I understand that considering religious beliefs is an extremely important part of this process for you. So tell me more about what you're thinking."

Self-Involving Responses

As we previously stated, self-involving responses are direct communications of the counselor's feelings about and reactions to the client in the here-and-now situation. Self-involving responses deal with the immediate counselor/client relationship; they are a type of feedback about how the relationship is personally affecting both the counselor and client(s). As such, they can be highly intimate responses. Self-involving responses can be particularly helpful when there are issues that are preventing the client from disclosing deeper feelings and thoughts; they are an effective way to enhance your genuineness, likability, and trustworthiness, and they can reduce client anxiety (Egan, 1994; McCarthy, 1982).

Counselor-Client Situations that May Prompt Self-Involving Responses

McCarthy and Oakes (1998) list several issues that can occur in mental health counseling and for which self-involving responses may be effective. We consider the following themes to be most relevant for genetic counseling:

Session is losing direction. "I feel as if we're getting off track here. Can we stop for a minute and see if we can figure out what's going on?"

Tension exists between client and counselor. Anxiety is a very contagious emotion; that is, it is easy to become anxious when you are with an anxious client. First, recognize the anxiety, second remain calm, and then talk about it in a composed and accepting manner. For example, you could say, "I'm feeling a little tense, and I wonder if you feel that way, too? Can we talk about what's going on to cause this discomfort?" Some signs that the client may be tense include behaving dependently, seeking continual reassurance from you, repeating the same questions, shifting the topic when you raise sensitive issues, making jokes, and frequently interrupting you.

Trust has not developed. "I'm concerned that you seem reluctant to talk with me."

There are cultural differences. "I'm concerned that because we are from different cultures, I might unintentionally say something offensive. Please tell me if that happens or if there are other things I can do to help you."

There are conflicting agendas. "I feel uncomfortable telling you what I would do in your situation because it seems like you want my permission to make this decision."

There are counterdependency issues. "It seems to me that we're working at cross purposes. I want you to hear all of this information, and you want to shut it all out."

You have just given the client bad news. "I'm sorry. I know this is not what you wanted to hear." This is a time when there's nothing you can say that seems to be adequate for the situation. Let your client regain some composure and then ask, "Is there something I can do right now?"

You sense the client is under pressure to decide. "I'm concerned that you may be rushing into a decision before you are ready."

The client appears to be angry. "I'm sensing that you may be angry, and I'm concerned that it's getting in the way of our talking right now."

Some Cautions About Using Self-Involving Responses

Self-involving responses are risky because you must put your feelings on the line. This requires you to first be aware of what you are feeling and then to be comfortable enough with your reactions to discuss them with your clients. Clients may react negatively to your feelings, or they may ignore them because they are uncomfortable discussing emotions. You may need to redirect clients to your self-involving statement if they move on without commenting (e.g., "I wonder if I could ask how you are feeling about what I just said").

You also need to remember that some members of certain cultural groups (e.g., Chinese clients) may not be comfortable discussing feelings directly. In this situation, you might try discussing feelings indirectly (e.g., "Some individuals might feel reluctant talking to a genetic counselor because they would not trust her").

Examples of Genetic Counselor Self-Involving Responses

Here are several examples of the types of self-involving responses that experienced genetic counselors have made during sessions:

- "I am feeling like you are angry at something. Is there anything I can do to help you?"
- "I get the feeling that there is something I'm not understanding. Can you tell me what's upsetting you?"
- "I hope that you have found this information helpful."
- "I am sorry that we can't be more certain about your risk."
- "I am concerned about you."
- "I can tell that this conversation is making you very uncomfortable. I know that you've been given a lot of information all at once."
- "I can see now that termination of your pregnancy is not an option for you, and I apologize if my bringing it up has made you uncomfortable. I just wanted to be sure that you were aware of all of your options."
- "I am terribly sorry to hear that your older son died." (This was not related to the consult; he did not die of a genetic condition. You should comment on issues that are extremely important to the client but are unrelated to the topic at hand).
- A client and her partner were in for follow-up genetic counseling to discuss a marker chromosome identified by an amniocentesis. Subsequent studies on the parents revealed that it was a maternally inherited marker. The father was angry, questioning the genetic counselor's credentials, and did not want to hear what information the counselor had to give them. The counselor said, "I sense that you are angry and not wanting to hear the information we have for you."
- The genetic counseling session involved a prenatal client and her partner. The ultrasound revealed a neural tube defect. Her partner was very hostile and distrustful. The counselor said, "I sense that you are having difficulty believing the information I am giving you."
- The client and her partner were seen for genetic counseling to discuss advanced maternal age. Her partner was reading a magazine during the session. The counselor said, "I feel somewhat uncomfortable with your reading a magazine. Your input is important in this discussion."
- "I'm worried that you are blaming yourself for your child's condition."
- "I'm very frustrated here because I want Mary to have the best possible chance in life—and I know that you want the same for her. She won't have that chance if she does not stay on her recommended diet."

Closing Comments

Advice, self-disclosure, and self-involving responses are powerful skills that do not occur as frequently as many of the other counseling skills we have discussed in this manual. Because they can be quite directive, we recommend that before you make a self-referent response you ask yourself two questions: "Whose needs are being served by my response?"; and "Could I achieve the same results with a less directive intervention?"

Self-referent responses are advanced helping skills. You can expect to become more comfortable and more effective using these skills after you have established solid attending, primary empathy, and questioning skills. Also, your self-reference skills should improve as you see more genetic counseling clients.

Class Activities

Activity 1: Think-Pair-Share Dyads

Students think about a time that someone gave them advice that was very ineffective. What was it like? How did the person present the advice? Now, think about a time when they received advice that was very effective. What was it like? How did the person present the advice? Have students discuss their experiences with a partner. Then the class discusses the following: What are ineffective ways to give advice? What are effective ways to give advice? The instructor summarizes their responses on the board or on an overhead.

Estimated time: 15 to 20 minutes.

Activity 2: Small Group Discussion

Students discuss in dyads or small groups the following question: What do you think we should sometimes, never, and always disclose to our clients? The instructor summarizes their comments in three columns on the board or overhead:

Never   Sometimes   Always

Estimated time: 10 to 15 minutes

Activity 3: Class Discussion

Students talk about what they think self-disclosure and self-involving responses are and how the two skills are similar to and different from each other. Students can respond to these questions in dyads in order to start the discussion.

Estimated time: 10 to 15 minutes.

Activity 4: Small Group Discussion

Students discuss the following questions: Would you ever stretch the truth or mislead a client in order to build the relationship? For example: You've gotten the preliminary results, and they don't look good, and the lab tells you not to say anything yet."

- Are there any advantages?

- Disadvantages?

Would you ever say you've had experience working with clients with a similar genetic condition when your experience is only one case?

- Are there any advantages?
- Disadvantages?

Estimated time: 10 to 15 minutes

Instructor Notes

- Try to draw out the following points during the discussion: (1) What does this do to trust? (2) What if you get caught in the lie? (3) The client will make all sorts of assumptions about what you understand and won't give you enough information to understand her or his situation. (4) How would students feel if a professional stretched the truth with them? (5) What does this say about their level of professionalism? (6) When you tell the truth you don't have to remember what you've said. (7) Honesty communicates counselor genuineness and respect.
- Next discuss what they could say instead. For example, "My experience with people with a similar situation is that they feel..." or "I've not had that experience, but if I did..."

Activity 5: Brainstorm: Responses to Client Requests for Self-Disclosure

We provided above (see Guidelines for Using Self-Disclosure) three responses to client requests for counselor self-disclosure. Respond to each of the following client questions in a nondisclosing way. Formulate three or four nondisclosing responses for each client question.

Client asks:

- Do you go to church?
- Have you had an amnio?
- Do you have children?
- Are you married?
- What do you think about abortion?
- Do you believe in God?
- Do you think it's wrong to wish my disabled child had never been born?
- Did you drink when you were pregnant?
- Was my doctor wrong to send my blood sample for further testing without asking me first?
- Do you think I'm being too hasty in my decision to terminate this pregnancy?

Estimated time: 20 to 30 minutes

Activity 6a: Low-Level Self-Reference Skills Model

The instructor and a volunteer genetic counseling client engage in a role-play in which the counselor demonstrates poor advice giving, self-disclosure, and self-involving behaviors. Students observe and take notes of examples of poor self-reference behaviors.

Estimated time: 10 minutes.

PROCESS

Students discuss their examples of poor behaviors and the impact of the counselor's poor skills on the client. The client can offer her or his impressions of the counselor's behaviors after the other students have made their comments. The instructor could divide the group of students and have half of them focus on counselor behaviors and half focus on client behaviors.

Estimated time: 15 minutes.

INSTRUCTOR NOTES

- One option is to continue with a role-play used earlier in the class so that the genetic counselor can more quickly move to self-reference skills.
- Students may have difficulty differentiating self-disclosure from self-involving responses. The instructor should assist them in categorizing the responses they observe in the role-play. One issue that may come up is that the genetic counselor's intended response may be perceived differently by the client and/or the observers.

Activity 6b: High-Level Self-Reference Skills Model

The instructor and the same volunteer repeat the same role-play, only this time the genetic counselor displays good self-reference skills. Students take notes of examples of good behaviors.

Estimated time: 15 minutes.

PROCESS

Students discuss their examples of good self-reference behaviors and the impact of the counselor's behaviors on the client. They also contrast this role-play to the previous role-play. The instructor could divide the group of students and have half of them focus on counselor behaviors and half focus on client behaviors.

Estimated time: 15 minutes.

INSTRUCTOR NOTE

- Students could work together in Think-Pair-Share dyads to identify examples of counselor self-reference and their impact on the client.

Activity 7: Role-Plays

Working in small groups, students volunteer to be client and genetic counselor. The students read their roles silently. Then the student playing the genetic counselor reads her or his role aloud. Next the students engage in a 10- to 15-minute role-play. If the student playing the genetic counselor gets stuck, stop the role-play and ask the group to brainstorm possible ways to handle the situation. Resume the role-play, so that the student can try out some of the group's suggestions. Ask the group to provide feedback to the counselor at the end of the role-play, and have a general discussion about how to handle each type of situation. Allow about 10 to 15 minutes to process each role-play.

CLIENT ROLE I

You are a 25-year-old woman who is discussing an abnormal amnio result with the genetic counselor. During this session, you say to the counselor, "Well, would you ever have an abortion?"

GENETIC COUNSELOR ROLE I

Your client is a 25-year-old woman who is discussing an abnormal amnio result with you. During this session she says. . .

CLIENT ROLE II

You are a 35-year-old who is talking about whether or not to pursue testing for Huntington disease. You should repeatedly ask the counselor for advice and each time the counselor gives you advice, you should say, "Yes, but. . ." and then go on to explain why the advice wouldn't work. During this session you say. . .

GENETIC COUNSELOR ROLE II

Your client is a 35-year-old who is discussing whether or not to pursue testing for Huntington disease. The client does not have a clear idea of what s/he wishes to do. During this session, the client says...

CLIENT ROLE III

You are from Taiwan. You are here to see the genetic counselor for prenatal counseling. Your English skills are not very good, and you have not been able to understand everything that the counselor is telling you. However, you nod your head and smile and pretend that you do. If the counselor brings this up with you, don't admit that you don't understand right away. After all, you don't want to be disrespectful to her.

Genetic Counselor Role III

Your client is from Taiwan. She is here to see you for prenatal counseling. Her English skills are not very good, and you do not believe that she has understood everything that you have told her. You say...

Client Role IV

You are a 38-year-old prenatal client who has been told by the counselor that there is a problem with the pregnancy. You cried when s/he gave you this information. You are embarrassed about crying and are also worried about what the genetic counselor must be thinking of you. Furthermore, you are afraid that you may lose control if you disclose any further.

Genetic Counselor Role IV

Your client is a 38-year-old prenatal client. You have just told her that there is a problem with the pregnancy. She cried when you gave her this information. You suspect that the client is embarrassed about crying and is also scared of what you think of her. Furthermore the client may be afraid of losing control if she discloses any further. What will you say to the client?

Client Role V

You are a 30-year-old woman whose father was recently diagnosed with Huntington disease. Your genetic counselor has just explained that there is a test that could reveal whether you also carry the gene for this condition. You say, "I just couldn't handle hearing that I have it, too. I'll take my chances. My husband and I are trying to have a baby, and I want to concentrate on that. Do you think I'm doing the right thing?"

Genetic Counselor Role V

Your client, a 30-year-old woman has a father who was recently diagnosed with Huntington disease. You have just explained that there is a test that could reveal whether she also carries the gene for this condition. She says...

Estimated time: 2 hours.

Activity 8: Advice, Self-Disclosure, and Self-Involving Triad Exercise

Students practice the three self-reference skills in 20- to 30-minute role-plays taking turns as counselor, client, and observer. Allow another 15 minutes of feedback after each role-play. The students should focus on using good helping behaviors.

Criteria for Evaluating Counselor Self-Reference Behaviors

Well-timed

Use sparingly

Concise

Intended to meet client goals

Appropriate depth/intimacy

Follow-up with empathy

INSTRUCTOR NOTE

- The observer or counselor may wish to stop the role-play if the counselor is obviously stuck. The triad can then engage in a brief conversation about the client's dynamics, issues, etc., to help the counselor. Then the counselor and client resume the role-play.

PROCESS

Discuss in the large group: How was it to do this exercise? What are you learning about self-reference behaviors in general? About yourself? What questions do you still have about these skills?

Estimated time: 1 1/2 to 2 hours.

Written Exercises

Exercise 1: Advice

Refer to Silver's (1991) eight beliefs of counselors who give advice. Provide one or two written disputations for each belief.

Exercise 2: Self-Disclosure

Write three or four paragraphs responding to these questions:

- What do clients see when they look at you?
- What might your physical characteristics mean to different clients?
- What do you think that your characteristics and actions might communicate?
- Select one of your personal characteristics that you think might affect a client negatively and write a brief dialogue (two or three interchanges) describing how you would address it with the client:

Counselor:

Client:

Counselor:

Client:

INSTRUCTOR NOTE

- Have each student write a few paragraphs on someone else in the class, or have them work on this in pairs. They may be able to tell each other a lot that isn't self-evident.

Exercise 3: Advice, Self-Disclosure, and Self-Involving Responses

Read each of the 14 client statements listed in Chapter 8, Exercise 2, and give one advice response, one self-disclosure response, and one self-involving response for each statement. Write your responses as if you were actually talking to the client. (Hint: You may have to infer more knowledge about the client in order to formulate your responses.)

EXAMPLE

For the example of the 40-year-old man at risk for Huntington disease cited in Chapter 8, Exercise 2:

Advice: "Perhaps you should reconsider what having the testing done will mean for you."

Self-disclosure: "I'd be very scared to learn that I had the gene. I wonder if that's how you are feeling, too?"

Self-Involving: "I'm concerned that you may be pursuing testing before you are ready."

Exercise 4: Role-Play

Engage in a 45-minute role-play of a genetic counseling session with a volunteer from outside the class. During the role-play focus on all of the helping skills you've learned so far. Try to include at least one advice, one self-disclosure, and one self-involving response. Audiotape the role-play. Next transcribe the role-play and critique your work. Give the tape, transcript, and self-critique to the instructor who will provide feedback. Use the following method for transcribing the session:

Counselor	Client	Self-Critique	Instructor
Key phrases of dialogue	Key phrases	Comment on your own response	Provides feedback on your responses

Annotated Bibliography

Diller, L. H. (1986). On giving good advice successfully. Family Systems Medicine, 4, 78-90. [Offers suggestions for advice giving.]

Kessler, S. (1992). Psychological aspects of genetic counseling. VII. Thoughts on directiveness. Journal of Genetic Counseling, 1, 9-17.
[Discusses advantages and limitations of directiveness in genetic counseling.]

Kessler, S. (1997). Psychological aspects of genetic counseling. X. Advanced counseling techniques. Journal of Genetic Counseling, 6, 379-392.
[Provides specific examples from genetic counseling and points out certain considerations and cautions in using self-referent interventions.]

Lippman, A. (1999). Embodied knowledge and making sense of prenatal diagnosis. Journal of Genetic Counseling, 8, 255-274.

[Offers important perspectives for the prenatal session that may be relevant for other types of genetic counseling as well.]

McCarthy, P., Oakes, L. (1998). Blank screen or open book? A reminder about balancing self-disclosure in psychotherapy. Voices: The Art and Science of Psychotherapy, 34, 60-68.

[Offers practical suggestions for counselor use of self-disclosure and discusses strengths and limitations of this type of counselor intervention.]

Silver, E. (1991). Should I give advice? A systemic view. Journal of Family Therapy, 13, 295-309.

[Provides a thoughtful discussion of the advantages and disadvantages of advice.]

Vontress, C. E. (1988). Social class influences on counseling. In: Hayes, R., Aubrey, R., eds. New directions for counseling and human development. Denver, CO: Love Publishing.

[Discusses possible cultural differences in desire for and use of self-disclosure.]

Weil, J. (2000). Psychosocial genetic counseling. New York: Oxford University Press.

[Discusses counselor use of self, and issues pertaining to nondirectiveness in genetic counseling.]

CHAPTER 11

Behaving Ethically

Learning Objectives

1. Examine one's motives for being a helper and how they influence practice.

2. Describe principles that guide ethical behavior of students and health care professionals.

3. Recognize several ethical challenges that confront counselors who see clients with genetic concerns.

4. Identify genetic challenges to informed consent and to maintaining confidentiality.

5. Apply ethical principles and models to cases involving genetic concerns.

Acting as a professional means responsibly addressing clients' needs and expectations. This requires first an understanding of yourself and the personal characteristics that you bring with you into genetic counseling sessions. For instance, examining your motives for being a genetic counselor can help you to identify the strengths you bring to clinical situations as well as shed light on situations where your needs or values might impede your ability to provide adequate care for clients.

This chapter addresses motivations for being a counselor, describes several ethical principles and models for ethical decision making, and identifies some of the major ethical and professional challenges you may encounter and the resources you can call upon to meet these challenges.

Genetic Counselor Motives

You have probably been asked by a number of people why you want to become a genetic counselor. What do you usually say? Your response to this question contains clues about your motives for becoming a genetic counselor. What are motives? They are desires, wishes, needs, or wants that direct most of our behavior. You probably have several motives for becoming a genetic counselor (Danish et al, 1980; Hill and O'Brien, 1999). These may include the need to feel a sense of accomplishment, the need for stimulation (intellectual, emotional, etc.), the need to have hope, the need to have fun, the need to have an existential purpose in life (Cavanagh, 1990), the need to help others, the need to feel powerful and in control, the need to feel competent, the need to be altruistic, the need for security (financial, social, etc.), the need to be liked, and the need to be respected. This is not an exhaustive list. You may have other motives as well.

Do you ever worry that you might have the wrong motives for becoming a genetic counselor? In our opinion, motives, in and of themselves, are neither right nor wrong, good nor bad. Rather, *how* and *when* we try *to* satisfy our needs may lead to positive or negative outcomes. For example, you may be pursuing a career in genetic counseling because it allows you to feel competent or capable. One positive aspect of this motive is that it will likely prompt you to continually build your skills and knowledge. On the other hand, if your desire to be competent is excessive, that is, you feel that you have to be successful with every client, the motive may drive you to try to get clients and supervisors to tell you that you did a great job even when you did not. Similarly, if you desire to be liked, one positive aspect is that you will probably be warm, non-threatening, and encouraging with your clients. But an excessive need to be liked could also lead you to avoid confronting clients and/or keeping clients from expressing any negative emotions toward you (e.g., anger). It is important that you recognize and periodically review your motives so that you can gauge their impact on your clinical practice. If you are aware of your motives, they will be less likely to negatively impact your clinical work (Danish et al, 1980). Understanding your motives is a continual process; you should periodically review them, as they will change over time and with experience (Danish et al, 1980).

One useful structure for thinking about different types of motives is Maslow's needs theory (Danish et al, 1980). Maslow grouped needs into five types:

- Physiological needs: basic survival needs such as food, air, and water.
- Safety needs: security needs such as shelter, an income, protection from threat, etc.
- Belongingness and love needs: love, affection, and acceptance.
- Self-esteem needs: personal worth and competence needs such as self-respect, recognition, and achievement.
- Self-actualization needs: the fullest expression of one's self through needs such as variety, challenge, and fulfillment.

Culture, Motives, and Values

Motives

Our motives are strongly grounded in our cultural values. Davis and Voegtle (1994) identify four major cultural settings or affiliations that shape our values:

RELIGIOUS AFFILIATION

- Affiliation of family of origin
- Religious group with which you identify
- Values or practices associated with the group with which you identify

SOCIOECONOMIC CLASS

- Socioeconomic class of family of origin (family into which you were born)
- Your own socioeconomic class
- Values or practices associated with the group in which you place yourself

ETHNIC GROUP

(Note: Ethnic group may overlap with religious affiliation.)

- Ethnic group(s) of family of origin
- Ethnic group(s) with which you identify
- Values or practices associated with the group with which you identify

OTHER GROUP IDENTIFICATIONS

(Note: Examples might include a community or neighborhood group, social action group, or group whose members share special interests such as music or sports).

- Other group identifications of family of origin
- Other groups with which you identify
- Values or practices associated with the groups with which you identify

We tend to think of cultural issues as those belonging to others. However, we all have cultures and therefore bring our unique cultural issues to any relationship. An important source of values for a majority of genetic counselors and counselors-in-training is the influence of Western cultural values. Consider the extent to which the following values, identified by Davis and Voegtle (1994), affect your expectations of yourself and others:

- Achievement and success: Especially financial success and recognition for one's achievement.
- Activity and work: Disciplined, productive activity is a worthy end in itself.
- Humanitarian mores: Emphasizes sympathy for the underdog and assisting those in need.
- Moral orientation: Life events are judged in terms of right and wrong.
- Efficiency and practicality: The practical value of getting things done is emphasized.
- Progress: There is an optimistic belief that, over time, life for all is getting better.

- Material comfort: Emphasizes the "good life," which includes conspicuous consumption of goods.
- Equality: There is a constant avowal of one's commitment to equality of opportunity.
- Freedom: There is a very strong belief in individual freedom and opposition to constraints on freedoms.
- External conformity: Uniformity of dress, housing, verbal expression, manners, and political ideas is valued. Deviance is discouraged.
- Science and secular rationality: Science is the means to gain knowledge and mastery.
- Nationalism and patriotism: There is a strong sense of loyalty to one's country.
- Democracy: There is a belief that everyone should have a voice in the country's political destiny.
- Individual personality: Every individual should be independent, responsible, and self-respecting. The group should not take priority over the individual.
- Racism and related group superiority: Stresses differential appraisal of racial, religious, and ethnic groups.

Look over this list and ask yourself which of these values you embrace and how they are associated with the settings that have contributed to your growth and development.

Values

The National Society of Genetic Counselors (NSGC) states its primary professional values in its 1992 Code of Ethics, found in Appendix A. It is important for all professionals to be able to identify their personal values, those of their profession, and those of the organizations in which they work. Knowing these values can help you decide what is an appropriate response in a given situation, and identify where there are conflicting expectations that you must consider when choosing how to respond. You may feel that your only obligation ought to be meeting the needs of your client. On the other hand, you may feel that other family members could benefit from knowing information about your client's genetic susceptibilities.

Ethics and Professional Challenges in Genetic Counseling

Ethical challenges are challenges to our personal and professional values. They are part of everyday practice for genetic counselors. Genetic counseling is a health care, counseling, and education profession. Since genetic counselors wear many hats, you will need to clearly define what is at stake when encountering ethical challenges and to identify strategies to address them

(Bower et al, 2002). The NSGC Code of Ethics (Appendix A) provides an overview of counselors' obligations to clients, society, and to themselves. Counselors will face some ethical dilemmas as a result of these multiple, and at times conflicting, obligations. In some instances, what is best for a family or for the broader society is not what is best for an individual client. Honoring one principle (such as respecting patient autonomy) might mean ignoring other obligations (like acting fairly). Value conflicts occur when counselors experience the tension of competing values, or when client expectations and counselor values conflict (McCarthy Veach et al, 2001). In such instances, you will need to select the principle(s) most important to a particular situation to guide your actions.

In the following section we describe six ethical principles that might be used as a basis for decision making. Then we focus specifically on informed consent and confidentiality because genetic counselors have reported that they frequently encounter these two challenges (Bower et al, 2002; McCarthy Veach et al, 2001). We cite case examples from health care professionals who serve clients with genetic concerns.

Guiding Ethical Principles for Health Professionals

Beauchamp and Childress (1994) have described principles from moral philosophy as a guide for health professional behaviors. These principles are based on respect for all persons as intrinsically valuable.

Respect for Client Autonomy

This principle, which focuses on the client's right to self-governance, has been a guiding value in health care since the late 1960s and early 1970s. Today, in nearly every state in the United States, a Patients' Bill of Rights legally declares that clients have a right to accept or to refuse any medical treatment. Enhancing client autonomy is a major goal of genetic counseling. The NSGC Code of Ethics says that counselors "enable their clients to make informed independent decisions, free of coercion, by providing or illuminating the necessary facts and clarifying the alternatives and anticipated consequences."

Strategies for enhancing autonomy include providing informed consent and acting in a nondirective way to ensure that clients' values, and not your values, determine the clients' decisions. (Bartels et al, 1997). Nondirectiveness has been a guiding norm for genetic counselors since the early days of the profession (Bartels et al, 1997). Although counselors generally believe that behaving in a nondirective way is a primary professional obligation, this belief has come under scrutiny in recent years. It is important to note in what circumstances you are directive or nondirective. Bartels et al (1997) investigated counselor descriptions of directive behaviors and found that there is a distinction between directing the process and directing the outcome of genetic counseling. Process refers to conducting a genetic counseling session in ways that benefit the clients you see. For instance, you are responsible for orienting the client to the session (e.g., describing the format, purpose, etc.). You also may need to help clients clarify the meaning of genetic information in their lives and help them identify their own values and decision strategies. And, if a genetic test is needed to accomplish a client's objective, you should recommend that a test be done (Bartels et al, 1997).

Directing the outcome means that you influence a client to act in concert with your values. This kind of influence violates the spirit of respecting the client's right to self-governance. In a situation where you believe that you must tell a client what to do, for example about whether to terminate a pregnancy, you must carefully consider your reasons for taking this step. You might ask whether you are responding to the client's needs/values or to your own needs/values. Situations where you direct the outcome should be the exception and not the rule in your counseling practice. Acting nondirectively does not mean that you cannot make suggestions or recommendations, however. Clients who see genetic counselors usually expect the counselor to have expertise that could be helpful to them.

Counselors committed to nondirectiveness sometimes question whether they must always do what the client prefers, and whether they must sacrifice their own values in the name of enhancing patient autonomy. The answer to that question is no. As a professional, you also have an obligation to abide by your personal and professional values. Being clear about your own values and motivations, as well as client motivations, can help to clarify where value conflicts may occur. The Code of Ethics indicates that counselors can refer to another professional when they cannot support the action a client chooses. This may be an appropriate response in situations where you feel that providing assistance to a client would mean supporting an action that you regard as morally wrong. In most situations, good communication will allow clients and counselors to understand one another and respect each person's right to act in accordance with his or her own values.

Nonmaleficence

Nonmaleficence means doing no harm; it is considered by many to be a bottom-line moral principle. However, with new technologies and the ability to provide information about the likelihood of illness in the future, counselors often find that the potential for harm coincides with providing benefit. For example, sharing susceptibility information can support clients who want to plan for a future situation. At the same time, however, this information can create stress and sometimes even major crises in the lives of the people to whom the information is given. Since we can't guarantee that we will not inflict harm, a more practical interpretation of nonmaleficence is to consider permitting harm only when harm is unavoidable, and ensuring that there is a corresponding benefit. An example might be encouraging a client to participate in genetic research only when s/he understands fully the possible risks as well as the benefits to participation.

It is important to keep in mind that harm can extend beyond the physical realm to include emotional and financial harm and harm to one's reputation or integrity. For instance, you might consider harm to a marital relationship, and even harm to employment or insurance status, as you assist clients with making decisions about genetic testing (Billings et al, 1992). Part of honoring autonomy and preventing harm is meeting a professional duty to provide informed consent about genetic tests. Given relevant information, clients will be more prepared to assess for themselves the harms and benefits of genetic testing or consequences of sharing test results. Preventing harm requires that you have a lifelong commitment to ensuring client safety, to keeping current with respect to professional standards and policies, to practicing within these standards, and to maintaining your competency to practice.

Beneficence

Beneficence as a principle to guide health care professional actions asks that we help others "further their important and legitimate interests" (Beauchamp and Childress, 1994, p. 260). Historically, beneficence was the major action guide for professionals. Section II of the Code of Ethics for Genetic Counselors asserts, "The primary concern of genetic counselors is the interest of their clients."

For health care professionals, beneficence usually means considering a client's medical best interests. For example, you might believe that most people would benefit by knowing whether they are highly susceptible to a familial cancer. Therefore, from your point of view, participating in genetic testing would be in their best interest. However, some clients, or their family members, will refuse to participate in testing because they fear that they could not handle receiving a positive test result. In this situation, beneficence, acting in a client's best interest, is challenged by duty to respect client autonomy. We act paternalistically or parentally when we assume that we are in a better position than the client's family to determine what is in their best interest. The NSGC Code of Ethics challenges paternalism by advocating respect for patient autonomy so that clients can act in accordance with their own values, even when you might disagree with their decisions. In the case above, you may want to first ensure that the client understands the risks and benefits of both testing and not testing. If you find that the client is making an informed, autonomous choice, your primary obligation would be to respect her decision.

Justice/Fairness

Justice is a principle that concerns the equitable distribution of burdens and benefits of health care. It includes preventing discrimination with respect to access to services. The NSGC Code of Ethics addresses nondiscrimination by saying that counselors "equally serve all who seek services and respect their clients' beliefs, cultural traditions, inclinations, circumstances, and feelings." In terms of counselors and societal obligations, the Code of Ethics invites counselors to "prevent discrimination on the basis of race, sex, sexual orientation, age, religion, genetic status, or socioeconomic status." Justice is a major moral challenge within the U.S. health care system. More than 40,000,000 people have no health care coverage. Medical indigence is a major cause of morbidity and mortality.

Many questions remain unresolved with respect to the types of genetic services that ought to be offered, what will be paid for, and how to balance cost with efficacy. For instance, you may consider, as other counselors sometimes do, whether you would offer a genetic test that insurance does not cover and the client could not afford. Most counselors would tell people about all options. You may, in that case, find that you need to seek funding and advocate for financial coverage.

Acting justly also means equally serving all who seek services. Wang and Marsh (1992) said that people may not be equally served if we rely only on the standard ways of operating that are part of the Western biomedical system when we work with people who are culturally or linguistically diverse. The concepts of patient autonomy, informed consent, and nondirective counseling may be inappropriate and may compromise cultural integrity as well as the delivery of effective genetic services. "Any discussion of Asian values must begin with the family as the fundamental social unit" (Wang and Marsh, 1992, p. 984).

Fidelity and Veracity

Fidelity and veracity are principles that support client autonomy. Fidelity, or faithfulness, includes promise-keeping, meeting contracts, and being trustworthy. Client-professional relationships are founded on trust and confidence. Thus, a genetic counselor is a trustee for clients' medical welfare (Beauchamp and Childress, 1994). The obligation to be faithful presumes that professionals have a duty to care for their clients. Clients generally assume that their interests are a primary concern of the counselor's, and professionals generally assume an exclusive focus on a client's best interest. McCarthy Veach et al (2001) found that counselors encounter challenges to honoring a client's wishes when they reflect on the impact that genetic information could have on other family members. Fidelity also challenges counselors to watch what they promise, and to let people know the limits and consequences of the genetic information that will be given. Concerning what you promise to clients, remember that there are mandatory reporting requirements related to illegal drug use and to child abuse and neglect. Therefore, you should inform clients that in such situations maintaining confidentiality would not be possible. With respect to consequences of genetic information, you will often give clients prenatal and presymptomatic test information and will need to tell them what that information could mean for future situations such as jobs and health insurance.

Veracity, or truthfulness, concerns your duty to tell the truth and not to lie or deceive clients. Truth telling supports client autonomy because clients can't be empowered to make decisions without accurate information. Veracity is closely related to fidelity and to sharing accurate information in the informed consent process.

Comments About Ethical Principles

Ethical principles can serve as guidelines for addressing ethical dilemmas. They provide a language that you can use to think about and to discuss challenging situations with colleagues. You will find, however, that an awareness of these principles does not automatically provide the answer to ethical questions. In fact, ethical dilemmas often consist of competing obligations, that is, a desire to do two mutually incompatible things. In some situations, you will want to maintain an individual's confidentiality, but will also be concerned with harm to family members if genetic information is not disclosed. For instance, you might see a parent affected with familial adenomatous polyposis (FAP) who refuses to disclose that fact to his or her children. Your concern might be that the child of an affected parent will not have information about the importance of screening that could prevent the occurrence of this cancer. In instances such as this, you must weigh the relevant principles, consult with other professionals, and choose the most important principle as a basis for action.

Bower et al (2002) found that genetic counselors draw upon many resources to help them sort out what to do when facing ethical dilemmas. Genetic counselors use a varied list of strategies:

- Further discussion with patient: to clarify possible implications of test results, the patient and/or family situation, and for policies regarding testing.
- Consulting with a health professional: consulting with genetic counselors, other health care professionals (depending on the expertise needed), and an institutional ethics committee for help with defining and resolving ethical challenges.
- Referral to a professional: recommending that the client seek assistance from another professional (see Chapter 6 on referrals).
- Informing/educating health professionals: addressing situations where other professionals have made errors, for example, erroneously ordering or interpreting test results; or, more generally, providing health professional education.
- Defer to preestablished rules or guidelines: following health care facility policies, professional guidelines [e.g., from NSGC or the American Society of Human Genetics (ASHG)], and consent policies.
- Advocate for the client: appealing to third parties to get medical care, genetic testing, or reimbursement.
- Withholding information: this strategy usually relates to situations involving requests for information from third parties. The counselor refrains from disclosing information, believing that it would be detrimental to the client.
- Disregarding personal beliefs and biases: taking a nondirective course by not expressing one's disagreement with the client's decision.
- Determine boundaries within a family: when facing conflicting duties to family members, clarifying to whom one's professional obligation lies, with

whom information should be shared, and which family member is responsible for decisions.

It is important, both while still a student and when working in a new situation, to learn about the available resources within the facility in which you practice. These include peer resources, supervisory resources, legal resources, ethical resources, and referral resources. It is especially important (particularly as a student, and in early practice) to have both support and consultation available. Peers tend to be the first recourse when genetic counselors meet a challenging situation. Sharing common experiences is important for reassurance, even when you have decided what is the best action in a particular situation.

Ethics committees exist in most health care facilities in the United States. Additionally, the NSGC has a committee that is available to assist with ethical dilemmas. Contact the NSGC at Web site http://www.nsgc.org/. The ethics committee is a subcommittee of the professional issues standing committee and can be directly connected to through the link http://www.nsgc.org/about_chairs.asp. These consultants provide recommendations, not imperatives, about the avenues available.

MORAL Model for Ethical Decisions in Clinical Situations

Patricia Crisham (1985) created the MORAL model for clinical decision making. This model is similar to the Danish and D'Augelli (1983) model for rational decision-making described in Chapter 7. However, Crisham explicitly identifies ethical principles as relevant factors. The moral model includes a grid shown in Table 11.1, for identifying the values that you wish to honor in making a particular decision, and the practical considerations that you need to take into account. Values might include the ethical principles previously described in this chapter, or they might include another explicit goal such as maximizing coping ability or avoiding escalation of family conflict. Practical considerations may include legal issues, time constraints, reimbursement concerns, and other factors that influence decisions in a particular work setting.

Prior to decision making, you need to describe the problem and the participants, including who makes the decision and what ethical value or principle is at stake. Crisham's (1985) decision-making steps spell out the acronym MORAL:

- Massage the dilemma: This process includes recognizing whose interests are involved in a conflict and defining the dilemma from their perspectives. The client's dilemma might result from a sense of conflicting loyalties. To formulate a goal for decision making, consider beginning with a sentence, "I would like to act in such a way that..."

- Outline options: List all alternatives in the column on the left side of the grid. In this stage it is important to consider as many options as possible. Brainstorming with other students or colleagues may reveal possibilities that you would miss by completing the grid alone.

TABLE 11.1. Moral model for ethical decision-making grid

Options	Values	Practical considerations

- Review criteria and resolve: Criteria are the values and practical considerations that you have identified at the top of the grid. You can go through each option, placing a plus where the criterion is met, or a minus in each column where that criterion is violated (Crisham, 1985). You may decide in this process that one criterion is really more important than the others. In that case, you would give the criterion more weight. Looking over the grid you will find that some alternatives meet more of your defined criteria than others do, and thus are more viable options for action.

- Affirm position and act: Now that you have done the moral analysis and have decided, based on that analysis, what you will do, you will need to consider a strategy for acting on that moral commitment. While anticipating any obstacles to action, it is important to keep the goal in mind.

- Look back: After taking action, you consider how successful you were and what worked and didn't work in the analysis of the ethical dilemma and the action taken. In this process, you will learn what works for you and what pieces of the model you will take with you into your daily practice of counseling and ethical decision making.

As you gain more experience, you will recognize regularly occurring ethical situations. Although you will still experience ethical dilemmas, you will become more comfortable knowing that you have the tools and resources to address them.

Two Ethical Challenges Encountered by Genetic Counselors

Providing informed consent and maintaining client confidentiality are challenges that occur frequently in genetic counseling situations (Bower et al, 2002; McCarthy Veach et al, 2001). Both of these challenges involve the principle of respecting client autonomy.

Informed Consent

The Patients' Bill of Rights in nearly every state of the U.S. requires that clients be told about benefits, risks, and alternatives to any treatment offered, so that they have sufficient information to make informed decisions. The purpose of providing consent is to enhance patient autonomy. Ideally, the counselor-client relationship is a collaborative one in which the counselor attempts to understand client expectations and explains how the counseling session will proceed. From this sharing, counselors can determine what information is material to client interests in this exchange (Geller and Holtzman, 1991).

Informed consent was the most prevalent issue in our study of ethical and professional challenges for primary care providers (McCarthy Veach et al, 2001). We found that several ethical challenges relate to this broad topic:

- Problems providing relevant and adequate information, including difficulties due to time constraints, the fact that it is not possible to convey every possibility, and professional biases that determine what information is relevant in a given situation.
- Client may not be capable of giving consent in situations where the client lacks legal competency (e.g., is a minor) or lacks the ability to reason through a problem (e.g., cognitively impaired).
- Client fails to understand, due to defenses or presumptions that lead to distortion of information.
- Coercion of the client in situations where family members or health care professionals influence a client to have a test or to act in a particular way.
- Client did not decide upon or authorize a genetic service that s/he received.
- Client misperceptions because the client was misinformed by the media, other health care professionals, etc.
- Client is unwilling to hear the information.

One key to providing adequate informed consent is to determine early in your relationship with a client what s/he expects and wants from genetic counseling. With this understanding you share relevant information about the anticipated benefits, uncertainties, and risks that come with genetic counseling and testing. For instance, when a patient wants to learn about her susceptibility to familial breast cancer you would encourage her to consider how testing will impact her life. She will need to know, for example, that having a positive gene test is not a 100% prediction that she will get breast cancer; furthermore, having a negative result does not eliminate the normal population risk of breast cancer. The client may or may not anticipate any risks of insurance or employment discrimination based on testing; you may need to explore the extent of such risks with her.

Jacobson et al (2001) surveyed prenatal genetic counselors about how they inform clients about the process of genetic counseling. They suggested that prenatal counselors should inform clients in writing about the genetic counseling session itself. Their sample of counselors generally agreed that the following content in is important for providing informed consent to prenatal clients:

- Description of services (e.g., length of the session, general format)
- Staff credentials (including whether one is a student under supervision)
- Risks and benefits of prenatal genetic counseling services
- Limits to confidentiality
- Alternatives to genetic counseling and to prenatal testing, especially the right to decline services
- Nonpaternity issues
- Due process: who clients can contact and how if they have concerns, complaints, etc. about the genetic counseling that they receive

• Patient/counselor signatures

Informed consent supports patient autonomy. Although no one can be perfectly prepared to make any medical decision, your efforts to share relevant information in a comprehensible way will enhance clients' abilities to understand the likely impact of genetics in their lives.

Confidentiality

Confidentiality refers to the protection of personal client information. Most clients assume that information given to health care professionals will be kept confidential and will be used only in the service of their own health care. However, in our study of ethical challenges (McCarthy Veach et al, 2001), we identified several challenges to confidentiality that you are likely to encounter.

Confidentiality within the family is a major concern since genetic information often has implications for other family members, as well as for the proband. In familial presymptomatic testing, many family members must provide blood for testing, but individual family members may or may not want test results. Ideally, you would ascertain from family members, prior to testing, the list of people with whom they are willing to share genetic information. The American Society of Human Genetics (ASHG) (1998) recommends that genetic professionals generally follow the legal and ethical principle of confidentiality. However, the ASHG cites situations that may permit disclosure: (1) when attempts to encourage the patient to tell family members have failed, (2) when serious and foreseeable harm is likely to occur, (3) when at-risk relatives are known, and (4) when either a disease is preventable or treatable or "medically accepted standards indicate that early monitoring will reduce the risk" (ASHG, 1998, p. 474). The professional seeing the patient and/or family must assess whether the harm that may result from failure to disclose information about genetic susceptibility outweighs the harm that may result from disclosure (ASHG, 1998).

Confidentiality within the health care system raises questions about how much information you should disclose to third parties such as insurance companies or other health practitioners. Today in health care facilities many auditors, including those from insurance companies and health plans, have access to the information included in a medical record. Where confidentiality is not possible, clients need to be told about potential consequences of having genetic test results included in their medical record. Additionally, a general rule of thumb is that medical record information will be disclosed only with written release from the client (Beauchamp and Childress, 1994).

A challenge specifically related to genetic information is whether to release test information from a family member who is deceased. Although, generally, this information is confidential unless a release of information has been signed, DNA of the family member who is affected with a genetic condition may be necessary for other family members to learn about their own susceptibility.

Finally, health care professionals are mandated by law to report some information, including danger or harm to self or others, including suspected child abuse and neglect, and, in some states, use of certain illegal drugs by pregnant women. Where these requirements exist, you will be in a position of deciding how to inform clients of these limits to their confidentiality. The informed consent process described by Jacobson et al (2001) may provide a useful vehicle for doing so.

Closing Comments

Genetic counselors frequently experience ethical dilemmas in their clinical practices. This chapter has described several frequent and important challenges that are likely to arise in your work with clients. We encourage you to discuss these challenges with your supervisors, peers, and other appropriate professionals.

As you continue in your career, we would also encourage you to continually assess your motivation for responding as you do, and to be aware of the ethical bases that guide your practice.

As you gain more experience, you will recognize regularly occurring ethical situations. Although you will still experience ethical dilemmas, you will become more comfortable knowing that you have the necessary tools and resources to address them.

Class Activities

Activity 1: Motives to Be a Genetic Counselor: Think-Pair-Share Dyads

• Maslow's hierarchy of needs describes five categories of needs: physiological needs, safety needs, belonging and love, self-esteem, and self-actualization. Can you think of one need from each of Maslow's categories that you could meet by becoming a genetic counselor? Take a few minutes to list the needs and to describe how being a genetic counselor could meet them. Consider which of these needs might be the most powerful motivator for you.

• Discuss with a classmate what needs might be a major reason that you chose to be a genetic counselor. Identify together the strengths and limitations this need brings to your practice.

Estimated time: 20 minutes.

Activity 2: Culture and Helpers, Part I: Discussion: Think-Pair-Share Dyads

Pairs of students discuss the following questions: (1) What is culture? (2) What is genetic counseling? (3) How does culture impact genetic counseling? (4) Which values or orientations described by Davis and Voegtle (1994) are most important to you? (5) How might they affect your work with clients? (6) What cultures/groups do you participate in? What are their prominent features? Of what are you the most proud? What do you appreciate the least?

Estimated time: 20 minutes.

Activity 3: Culture and Helpers, Part II: Discussion: Think-Pair-Share Dyads

Pairs of students discuss the following questions: (1) What kinds of differences in culture do I perceive in others? (2) How do I respond to the differences that I see in others? (3) What biases am I aware that I have?

Estimated time: 20 minutes.

PROCESS FOR ACTIVITIES 1 TO 3

Why do you think it is important to discuss motivations for being a counselor in relation to cultural differences? What did you learn from these exercises? What are your feelings about examining your cultural values, your motives, and cultural differences?

Estimated time: 10 to 15 minutes.

Activity 4: Addressing a Moral Dilemma Using the Analysis from Exercise 3: Think-Pair-Share Dyads

- What option did each select?
- What is your rationale for selecting that option?

After discussing these responses, students consider together what elements of the model explain the similarities, or differences, in their choices.

Estimated time: 45 minutes.

Activity 5: Structured Controversy Directions

The structured controversy exercise gives you an opportunity to try on and evaluate alternative perspectives. Follow the steps to address the situation below.

SITUATION

A 21-year-old client with an autosomal-dominant form of mental retardation became pregnant. The client's mother requested that the pregnancy be terminated, while the client wants to continue the pregnancy. You question the client's ability to comprehend the situation. You need to decide whether you agree with the mother's position or with the daughter's (Bower et al, 2002).

PROCESS

- Assign one person to be the timekeeper.
- Count off the participants; odd numbers will take one position, even numbers will take an alternate position.
- Initially the odd-numbered people will take the nondirective perspective

and the even numbered people will take the directive perspective.

Estimated time: 5 minutes.

ASSIGNED PERSPECTIVE

- Prepare presentation with partner(s) (20 minutes).
- Present arguments (5 minutes per side; 10 minutes total).
- Ask clarification questions when the opposite side finishes presenting (5 minutes total).
- General discussion (5 minutes).

Estimated time: 40 minutes.

REVERSE PERSPECTIVE

- Prepare presentation with partner(s) (10 minutes).
- Present arguments not included by alternative side (5 minutes).
- Present arguments (5 minutes per side).

Estimated time: 20 minutes.

OPEN DISCUSSION: DECISION MAKING

- Drop perspective.
- Seek and provide clarification, elaboration, justification, rationale.
- Summarize arguments.
- Reach conclusions.

Estimated time: 15 minutes.

REPORT PREPARATION

- Prepare a written response to the question (your group's conclusion, the data/arguments supporting your conclusion, the counterarguments to your conclusion, and the weaknesses of those counterarguments).

Estimated time: 20 minutes.

SMALL GROUPS DISCUSS CONCLUSIONS/RATIONALE WITH LARGE GROUP

Estimated time: 15 minutes.

Written Exercises

Exercise 1: Cultural Impact on Motives to Be a Genetic Counselor

Prepare a three- to four-page, double-spaced, word-processed paper discussing the following:

- How do you define culture?
- Did your personal definition of culture change as a result of the class discussion? If so, how?
- Describe your own ethnocultural background.
- Identify three of your personal motives (desires, needs) and discuss how they are potentially beneficial and potentially harmful in your practice of genetic counseling. The criterion for a beneficial motive is whether its consequences are likely to help rather than harm clients. The criterion for a harmful motive is the reverse. (Hint: Motives are not the same thing as traits. For example, it is not sufficient to say that one of your motives is that you are empathic. Why or how does this trait lead to a desire to be a genetic counselor?)
- What impact has your culture had on your motives to be a genetic counselor?

Exercise 2: Informed Consent

Consider the following questions:

- What three things might clients want to know about the purpose of genetic counseling?
- What might they want to know about the process and outcome of genetic counseling?
- What might they want to know about you (assume you are still in a student role)?
- What would you tell them about the limits to confidentiality?
- What would you want to know if you were the client?

Create a checklist of items you will discuss with all clients so that they are informed about genetic counseling. Next, compare you checklist with that of another student, and then create a list that includes your combined list of items to describe genetic counseling.

Exercise 3: Addressing an Ethical Dilemma

Using the MORAL model grid in Table 11.1, list all of the options you might have to define and address the ethical dilemma described in the case example below. At the top of the grid list the ethical principles (see list of ethical principles) you would like to meet in your response: "I want to act in such a way that..." Then list practical implications, such as the availability of a test, reimbursement, or legal concerns that also must be considered. Work through the MORAL model grid. Select the option you would choose, and the rationale, based on principles and practical considerations, for your decision.

Case Example*

Ms. W., a 28-year-old single woman, referred herself for genetic counseling

---

*From Maley, 1994.

because she was worried about the chance that she would develop a brain tumor, as her brothers had. With her assistance, Andrea, the genetic counselor, obtained records that indicated that the brothers had suffered from astrocytomas. Just before the second brother died, the diagnosis of neurofibromatosis type I (NF-I) was made. Since that time, one of his children has been diagnosed with NF-I. The older deceased brother had identical twin daughters, now 5 years old, who had also been diagnosed with NF-I. Ms. W. had one surviving brother and a sister who also wanted to know about their risks. The rest of the family history was remarkable only in that Ms. W.'s father had died of lung cancer at age 40. Ms. W. did not think anyone else in the family had NF-I. However, she admitted that her mother had quite a few birthmarks.

Andrea explained that determination of Ms. W.'s risk for carrying the gene for NF-I could be accomplished only by linkage, which required blood samples from several family members. The family members agreed to participate. Her mother agreed to have a blood sample drawn at her doctor's office and sent for testing but declined to be examined or to receive genetic counseling.

Andrea sent the blood samples to a lab for testing. The lab subsequently returned to Andrea a single report containing linkage data, haplotypes, and risk figures for all individuals. The affected nieces and nephew shared a haplotype with their grandmother, Ms. W.'s mother. Ms. W.'s sister also carried this haplotype. Ms. W. and her surviving brother had inherited a different haplotype from their mother. Based on this information, Andrea told Ms. W. that her risk of having the gene for NF-I was low.

When Andrea offered to discuss the results of the testing with Ms. W.'s mother, the older woman declined to hear about the results and told Andrea that she did not want information about her to be shared with anyone.

Ms. W.'s sister, a nurse, was understandably upset to hear that she shared a haplotype with her affected nephew and nieces. She asked to see the original test report from the lab.

Andrea believed that the mother's wishes should be respected, but was not sure what to do with the laboratory report or how to satisfy the sister's request to evaluate the information herself. She called the lab and asked if they could prepare separate reports. Laboratory personnel told her that because linkage is a family test, family members should understand that others might find out about their risks. Andrea was uncertain how to handle the original report, and how to document the results for each family member in his or her own file.

Annotated Bibliography

American Society of Human Genetics Social Issues Subcommittee on Familial Disclosure (1998). Professional disclosure of familial genetics information. American Journal of Human Genetics, 62, 474-498.
[Describes the ethical principles that ground confidentiality and conditions under which one might breach confidentiality.]

Bartels, D. M., LeRoy, B. S., McCarthy, P., Caplan, A. L. (1997). Nondirectiveness in genetic counseling: a survey of practitioners. American Journal of Medical Genetics, 72, 172-179. [Describes counselors' views of nondirectiveness; describes reasons for directiveness and discriminates directiveness in process and outcome of genetic counseling.]

Beauchamp, T. L., Childress, J. L. (1994). Principles of biomedical ethics (4th ed.). London: Oxford Press. [These authors describe ethical theories and principles, their rationale, and they apply ethical principles to health care issues.]

Billings, P. R., Kohn, M. A., de Cuevas, M., Beckwith, J., Alper, J. S., Natowicz, M. (1992). Discrimination as a consequence of genetic testing. American Journal of Human Genetics, 50, 476-482. [These authors collected anecdotes from people who experienced health care and employment discrimination.]

Bower, J. A., McCarthy Veach, P., LeRoy, B. S., Bartels, D. M. (2002). Ethical and professional challenges: a survey of counselors' experiences. Journal of Genetic Counseling, 11, 163-186. [Summarizes counselors' specific ethical challenges and how they are resolved.]

Crisham, P. (1985). MORAL: How can I do what's right? Nursing Management, 16(3), 42A-G. [Focuses on a model for ethical decision making that includes ethical principles and practical considerations.]

Danish, S. J., D'Augelli, A. R. (1983). Helping skills II: life development intervention. New York: Human Sciences Press. [Provides a detailed outline/guide for transforming client problems into positive goal statements, identifying roadblocks to goal attainment, and strategies for removing the roadblocks.]

Jacobson, G. M., McCarthy Veach, P., LeRoy, B. S. (2001). A survey of genetic counselor's uses of informed consent documents for prenatal counseling sessions. Journal of Genetic Counseling, 10, 3-24. [Describes the purposes of informed consent and compares consent processes in mental health and genetic counseling.]

Maley, J. A. (1994). An ethics casebook for genetic counselors. Wallingford, PA: NSGC. [Maley and an ad hoc committee of genetic counselors describe a model for thinking about ethical challenges. This book includes cases that counselors analyze using the Code of Ethics for Genetic Counselors.]

McCarthy Veach, P., Bartels, D. M., LeRoy, B. S. (2001). Ethical and professional challenges posed by patients with genetic concerns: a report of focus group discussions with genetic counselors, physicians, and nurses. Journal of Genetic Counseling, 10, 97-119. [Describes 16 domains of ethical challenges identified by physicians, nurses, and genetic counselors who see clients with genetic concerns.]

Wang, V., Marsh, F. H. (1992). Ethical principles and cultural integrity in health care delivery: Asian ethnocultural perspectives in genetic services. Journal of Genetic Counseling, 1, 981-992. [Deals with dilemmas that stem from counseling those of Asian descent. Issues of collective vs. personal autonomy come up, as do issues of nondirectiveness vs. expectations of authority and different conceptions of normalcy.]

CHAPTER 12

Recognizing Your Limits: Transference, Countertransference, Stress, and Burnout

Learning Objectives

1. Define client transference.
2. Identify ways to respond to transference.
3. Define genetic counselor countertransference.
4. Identify strategies for managing countertransference.
5. Define genetic counselor stress and burnout.
6. Identify stress and burnout coping strategies.

To be an effective genetic counselor, you must be aware of issues that impact your relationship with clients. This chapter discusses two critical issues that affect genetic counseling relationships: (1) transference and counter transference, and (2) stress and burnout. Transference and countertransference are primarily (but not always) unconscious dynamics that emerge within the relationship itself, while stress and burnout are conditions that develop in the counselor and spill over into genetic counseling relationships.

Transference and Countertransference

The primary focus of this section is countertransference, a phenomenon in which your own needs and experiences can affect your clinical work. Sometimes countertransference occurs in response to client transference. Therefore, we begin by briefly discussing client transference.

Definition of Client Transference

Transference is an unconscious way that a client relates to the genetic counselor based on her or his history of relating to others (Djurdjinovic, 1998). It concerns how the client perceives the counselor and how the client behaves toward the counselor. For instance, clients project onto the counselor attitudes, roles, and expectations based on previous encounters with others. Transference is a client's misperception of the counselor that can occur from the first moment of contact and even in anticipation of the genetic counseling session. An important aspect of transference is that the client's feelings and reactions tend to be overreactions to the reality of the situation. Because it is an unconscious process, clients are unaware that they are experiencing transference (Cerney, 1985). It is normal to have transference (e.g., experiencing an immediate attraction to or dislike of someone for no apparent reason). Upon some reflection, however, you realize it's because the individual reminds you of a family member or friend. Transference tends to be stronger when the counseling relationship is longer and more in-depth. It may be positive (feelings of affection or dependency), or negative (feelings of hostility and aggression).

Transference commonly involves ambivalence toward those in authority, dependency, or affection. Clients may experience cultural transference, relating to you by transferring positive or negative feelings from prior experiences with individuals from your cultural group (Pedersen, 1995). Clients who have transference reactions toward you may perceive you in one of five ways (Watkins, 1983):

- Counselor as an ideal: You are the perfect individual. Client behaviors may include excessively complimenting you, and/or continually agreeing with everything that you say. Similarly aged clients, for instance, may identify with your seemingly perfect life.

- Counselor as seer: You have all of the right answers. Client behaviors may include repeatedly asking you what you would do. Keep in mind, however, that clients may have cultural values that lead them to place health care providers in a position of authority (e.g., some Asian clients). This is not transference.

- Counselor as nurturer: You are their source of strength. Client behaviors may include acting helpless, excessive crying and emotionality, and urgent requests/demands for solutions to their problems.

- Counselor as frustrator: You are the spoiler of their experience. Clients may be excessively defensive, have minimal to no self-disclosure, and inappropriately blame you for the bad news that you communicate.

- Counselor as nonentity: You are an inanimate figure with no feelings, wishes, or needs. Client behaviors may include topic shifting, talking nonstop, and dismissing/ignoring your interpretations and reactions. For instance, sometimes in prenatal genetic counseling with couples, one of the partners will read a newspaper during the session.

Remember that client transference is based on client misperceptions and results in overreactions to the reality of the situation.

Examples of Client Transference

- A client of advanced maternal age had an amniocentesis that revealed Down syndrome. The client terminated the pregnancy by induction of labor. The genetic counselor referred the client to a support group and provided some counseling over the phone. The client made an excessive number of phone calls to the counselor and wrote her a number of letters. She requested a follow-up session but would not come into the building, asking instead that the genetic counselor have lunch with her. None of the genetic counselor's suggestions or referrals to other health professionals would work. The client continued to be helpless and was not open to any suggestions.

- A client was seen for genetic counseling due to an abnormal ultrasound. Amniocentesis revealed trisomy 18. The genetic counselor called to give the client the results of the amnio, and the client acted as if she did not remember meeting with the counselor. The counselor gave the client some information about their last visit and then asked if the client remembered her. The client replied in a very hostile manner, "How could I forget you!" The client was defensive and would not disclose any feelings about the test results.

- A client was the mother of a 2 1/2 year old child newly diagnosed with Angelman syndrome. The child had a history of moderate to severe developmental delay, no speech and language, and seizures. The mother was a single parent and was managing a career and this child fairly well. She had, however, an exceptionally intense reaction to the diagnosis, especially the mental impairment component. She displayed excessive crying and emotionality and wanted the genetic counselor to be a nurturer. She called several times with urgent questions and had a need to go over the information repeatedly. She requested another appointment to go over the information yet again.

- A couple had an intensely negative reaction to a geneticist who was present during their counseling session in which a prenatal diagnosis of Down syndrome was given. They wrote a letter to the head of the hospital complaining bitterly about his lack of compassion, when in fact he had behaved appropriately and in a caring manner.

- A 33-year-old client was referred because her triple screen showed that her risk for Down syndrome was 1 in 44. The risk for Down syndrome in her last pregnancy was 1 in 180, but she reported that everything "turned out perfect." Although she had never met with a genetic counselor before, she assumed that the genetic counselor was a "spoiler" of her experience. She hardly listened during the session and kept saying, "Those blood tests are always wrong anyway." She even refused to have an ultrasound because she didn't want to be told any more "fake bad news."

- A fairly common transference situation involves clients who come to the clinic completely exhausted or enraged due to a difficult commute or difficulty parking at the agency. Although the travel experience is not without tribulations, the clients usually have transferred their concerns about undergoing a genetic consultation. They may later say that getting to the office was not that big of a deal; they were just nervous about the visit.

- A genetic counselor had several clients who seemed to become enraged regarding maternal serum alpha-fetoprotein (MSAFP) tests. The fact that the test is a screening test and not a diagnostic test was never explained to them. They arrived with the assumption that their child has spina bifida, and when counseled that most likely the dates of the pregnancy were incorrect or that they were having twins, they reacted angrily toward the person (usually their health care provider) who informed them that a blood test showed that the baby had spina bifida. In reflecting upon these situations, the genetic counselor stated that she probably has reacted with some counter-transference of her own by wanting to protect the providers and her relationships with them by explaining how the MSAFP test is marketed poorly.

Responding to Client Transference

There is no single way to respond to client transference. Depending on the particular situation and client, you might try one of the following strategies:

- Simply accept it. Handle transference feelings as you would any type of client feeling. Let the client express the feelings, and either take them back or continue to express them.

- Recognize it. Recognizing that transference may be happening allows you to understand it and to avoid overreacting to client distortions (Djurdjinovic, 1998). Clues that transference is happening include your feelings of confusion and discomfort and your belief that the client's behavior contains some amount of distortion and misperception (Djurdjinovic, 1998).

- Decide whether or not to address it. You should be careful about drawing client attention to transference because, "If the transference is confronted and the counselee is unaware of the distortion, the resulting confusion may elicit a more intense response from the counselee" (Djurdjinovic, 1998, p. 137). Given your time constraints and the scope of what you must accomplish, genetic counseling may not be the appropriate venue for dealing with transference. However, you might choose to gently address mild expressions of transference in order to decrease their impact on the session (Djurdjinovic, 1998). For example, a client says that she is angry that you cannot absolutely ensure a healthy child based on prenatal test results. You might respond, "I wonder if part of what you're angry about is that you think I'm giving you the runaround, just like the other medical professionals you've seen."

- Ask clarifying questions. For example, "It seems like you are unloading on me. I wonder why this is happening?" This is a prelude to interpretation (see Chapter 8), but first explores the client attitudes and gives the client the opportunity to do her or his own interpreting.

- Reflect transference feelings. For example, "You said that you don't want to discuss your infertility history because you think it may make me uncomfortable?" Further discussion might prompt the client to acknowledge that negative reactions from family members have made her wary of talking about it with anyone. You could then assure her that you can handle whatever she wishes to disclose.

- Interpret transference feelings directly. For example, "Sometimes when people feel they have been telling too much they get uncomfortable about their relationship with that person. Do you suppose this is happening here?"

While some authors believe that transference (and countertransference) is a part of every relationship, including the genetic counseling relationship (e.g., Kessler, 1992), others believe that its occurrence depends upon the depth of the relationship that forms and on the interplay of the personalities of the genetic counselor and client (Djurdjinovic, 1998). Whether or not transference occurs in all relationships, it is important to remember that not all client reactions are transference. Clients may be responding appropriately to the situation and to you (e.g., feeling annoyed if you are late to the session; feeling confused if your presentation of information is too complicated or rushed). However, if you have not made a mistake, or if their reaction is greatly exaggerated (e.g., being furious because you are a couple of minutes late), transference may be occurring.

Definition of Counselor Countertransference

Countertransference has been described as the same phenomenon as client transference, but in the opposite direction (Cerney, 1985). Countertransference refers to the genetic counselor's unconscious ways of relating to clients based on the counselor's history of relating to others (Djurdjinovic, 1998). Countertransference can include emotional reactions and projections toward clients that may not be particularly appropriate to the current genetic counseling relationship (Brammer and Shostrom, 1982; Cerney, 1985). For example, you might find yourself feeling angry with a client who decides to terminate a pregnancy for sex selection when you have struggled with infertility. Or, when a client who is about your age tells you that she has been diagnosed with ovarian cancer, you may find yourself asking questions because of your own anxiety — "Did you have any symptoms? How did you find out you had it?" Similar to transference, countertransference involves misperceptions (e.g., viewing a client as too dependent when he isn't) and overreactions (e.g., anger over some client behavior that most counselors would be able to take in stride).

Why does countertransference occur? You and your clients may be similar or different in a number of ways such as your values behaviors, attitudes, language, physical appearance, age, gender, etc. (Gartner et al, 1990; Watkins, 1985). These similarities and differences affect the ways in which you identify with the client. Countertransference can occur when you have extreme overidentification with a client (you perceive the client as "just like me"). It can also occur when you experience extreme disidentification (you perceive the client as "nothing like me") (Watkins, 1985). When you overidentify, you become all wrapped up in your client's situation and have difficulty distinguishing where the client's feelings stop and yours begin. For example, you may find yourself thinking about and feeling very involved with a client who is your gender, about your age, and who has muscular dystrophy. The more you perceive yourself as similar to a client, the greater the chance of overidentification. When you disidentify, you feel disconnected, disengaged, and possibly even become hostile and rejecting toward the client. For instance, consider a couple that wants to terminate a pregnancy because the woman is carrying twins, and they only want one child. You may consider their decision to be morally offensive and find yourself pulling away from them. The more you perceive yourself as dissimilar to a client, the greater the chance of disidentification.

Countertransference can have both negative and positive consequences. You should be particularly concerned about negative effects. One possible negative consequence is that it can interfere with your empathy for a client. You may learn something from a client that triggers your own experiences, and soon you have stopped listening to your client and are busy thinking and feeling about your own situation (Kessler, 1992). You may think that your thoughts and feelings are about your client, but they are actually (and usually unconsciously) about you. Additionally, your client's situation may reopen current or old hurts, and because this is painful, you may avoid exploring the client's feelings, especially if your typical coping style is to distance yourself emotionally (Kessler, 1992). One possible positive consequence of countertransference is that by triggering experiences from your past, your might have increased empathy for clients (to the extent that your experiences are similar to theirs). However, as already stated, there is always the risk that you will pay less attention to clients and instead impose your experience on them.

As with transference, you must be careful to distinguish countertransference from situations in which your reactions are a realistic response to your client and her or his behavior (Cerney, 1985). For example, it is realistic to feel sad when a client is grieving over a pregnancy loss. However, if you become quite distraught over your client's situation it might indicate that you have unresolved feelings about a past loss. It is natural to feel irritated at a client who lies about being at risk in order to get a prenatal test. But if you become very angry with this client, it might suggest experiences from your past where you felt manipulated or controlled by others (see self-involving responses in Chapter 10, for a more extensive discussion of realistic counselor reactions). It is also important to distinguish countertransference from your natural, empathic response to clients (Cerney, 1985). Empathy, as discussed in Chapters 4 and 8, involves the ability to experience client feelings as if they were your own, while maintaining enough distance to realize that they are not your feelings. It also involves the ability to listen to the client's story without imposing your own assumptions.

Types of Countertransference

Kessler (1992) describes two types of countertransference in genetic counseling:

- Projective identification: This occurs when you mistakenly believe that your feelings are your client's feelings. So, for instance, if you feel a great deal of discomfort, but you think this is what your client is feeling, you might encourage the client to focus on less distressing ideas or images. When this happens, you will only be able to have shallow empathy at best because you won't go deeply into feelings that are too upsetting to you. Projective identification also occurs whenever you have the misperception that you understand exactly what a client is going through because you have had the same or a similar experience.

- Associative countertransference: In associative countertransference your client's experience taps into your inner self and you begin to focus on your own thoughts, feelings, and sensations. As with projective identification, associative countertransference is triggered by your own past or current problems or situations that are similar to your client's. However, a major difference is that you do not believe that your feelings are the client's feelings. You know that they are your feelings. When you experience associative countertransference, you find yourself losing focus. Your attention shifts from the client to yourself. You may find yourself daydreaming about your situation and realize that you haven't really heard anything the client has said for a few seconds or longer. Associative countertransference can be quite common. As Kessler (1992) points out, genetic counselors must deal with clients who have similar difficulties and problems that they currently are experiencing or experienced in the past. He further states, "Bad things do happen to genetic counselors. But even if they do not, we are as vulnerable as the next person to experience loss and pain. Disappointment, loss, feelings of being rejected and misunderstood, of failure, embarrassment, hurt, and so on are ubiquitous human phenomena. No one is exempt" (p. 304).

Watkins (1985) identified four types of countertransference that can either occur occasionally, with some clients, or can be more pervasive, occurring with many clients:

- Overprotective countertransference: You regard some or most clients as childlike and in need of great care and protection, so you cushion the information you give ("This risk really isn't so bad"); or you qualify your interpretations ("I'm wondering if you might be afraid of making the decision. I'm probably off base with that. It was just a guess. Forget I even mentioned it"); or you don't allow clients to experience and express their painful feelings ("Everything will be OK. It will be fine") (Watkins, 1985). Other evidence that you may be overprotective includes talking in a low voice and using physical gestures such as patting clients on the back, hugging, or patting their hands, all of which can be perceived as infantilizing (Watkins, 1985). An additional aspect of overprotective countertransference is excessively worrying about a client, even to the point of obsession (dreaming about the client, looking for reasons to contact the client, etc.).

- Benign countertransference: This type of countertransference is often due to an intense need to be liked by clients or to a fear of strong client affect, especially anger (Watkins, 1985). To prevent being disliked or to avoid strong affect, you create an atmosphere that is the same across all clients and situations, one that is characterized by shallow exploration of emotions, by optimistic, cheerful interchanges, and by limited consideration of negative information or issues (Watkins, 1985). There may also be a lot of extraneous chitchat as you attempt to be more like a friend than a genetic counselor.

- Rejecting countertransference: Similar to overprotective countertransference, you may regard some or most clients as dependent and needy, but you react punitively, becoming aloof or cold, and behaving in ways that create distance between you, either because you fear the demands clients might place on you or you are afraid of being responsible for their welfare (Watkins, 1985). Examples of distancing behaviors are blunt explanations ("You know that your risk of having a child with a birth defect is high because you drank during your pregnancy"), and dismissive responses to client requests ("That's your decision. I'm not you"). The following two examples of rejecting countertransference involve genetic counseling student statements made during supervision: "My explanation of the genetic condition wouldn't have been so confusing if the client had just given me a chance to explain!"; and "I don't know where she got the idea that I wanted her to terminate her pregnancy... She was just looking for someone to tell her what to do." In the first example, it is important to note that the client did give the counselor the chance to explain, and in the second example, the client did not wish to be told what to do.

- Hostile countertransference: This type of countertransference occurs when you dislike something about your client (e.g., a mannerism, a physical characteristic, an attitude, a value), and in an attempt to be as unlike the client as possible, you try to distance yourself in both overt and covert ways (Watkins, 1985). You go even further than with rejecting countertransference, perhaps making harsh statements (e.g., "I already told you that I'm not the one making this decision!"; or, "Boy, you sure have gotten yourself into a mess with this nonpaternity situation"). Your attitude is that the client deserves what s/he is getting (Watkins, 1985). Hostile countertransference may be more common in counselors who are experiencing some degree of burnout. Perhaps they are working repeatedly with a client population that is very needy and for whom there is little room for change in the system (e.g., a medical assistance population, new immigrant populations).

Behaviors that May Indicate Countertransference

Counselor countertransference is sometimes a reaction to client transference. For example, some clients may expect you to be a nurturer, and their demands prompt you to engage in rejecting countertransference. In addition, countertransference may be prompted by a particular type of client (e.g., terminally ill or cognitively impaired) and/or by certain genetic counseling situations that "push your buttons" (e.g., sex selection; presymptomatic testing of minors) or it may be a more habitual type of reaction that you have toward all or most of your clients (e.g., distancing from client affect; being overly protective, etc.). Countertransference (and transference) can be very complicated to identify and resolve, especially since it occurs primarily at an unconscious level. You may have to carefully observe and explore your overt and covert behaviors to detect its occurrence. The following behaviors may indicate countertransference, especially if you exhibit more than one of them:

- Engaging in compulsive advice giving (Corey et al, 1984).
- Having unusually strong feelings toward a particular client (Martin, 2000).
- Having "rescuer" fantasies, that is, believing you will really be able to help a client, even when others have failed (Martin, 2000), or when there is no resolution for the situation (e.g., the client is terminally ill).
- Dreading a session or being overly eager for a session with a particular client (Martin, 2000).
- Feeling sleepy during the session (Martin, 2000).
- Avoiding client feelings, especially negative feelings directed toward you. You may avoid feelings by showing disapproval (frowning, interrupting, etc.), using fewer reflections, and/or giving excessive information (Martin, 2000).
- In a few instances, your clients' behaviors may provide you with a clue that you are experiencing countertransference (Cerney, 1985). For example, your client says, "You sound just like my mother..."; or "You look so upset! I'll be OK"; or "I know you want me to make a decision, but I just can't yet!" (One caution in interpreting these client comments as signs of your countertransference is that they may be due to the client's transference!).

This is not an exhaustive list of countertransference behaviors. Generally speaking, any behavior, thought, feeling, or attitude that is either out of character for you or is considered by others (e.g., your supervisor, the client) to be ineffective or inappropriate could signal countertransference.

Countertransference Examples

- The genetic counselor saw a family with whom she shared many similarities. She knew that she was overidentifying with this family, and the counseling (for her) was much more intense emotionally. The client was a 14-year-old healthy girl, the same age as the genetic counselor's daughter. The client's mother brought the client to clinic for carrier testing for adreno-leukodystrophy (ALD). The client had a brother who died at age 10 of the disease. The genetic counselor also had a 10-year-old son. The counselor felt very strongly that this client should not be tested, but she tried to present the pros and cons objectively. She felt a need and desire to protect this 14-year-old girl. The girl's mother was clearly still struggling with her grief and was very fragile. All of the mother's reasons for testing seemed to relate to her own needs rather than to the client's needs.

- In the early days of chorionic villus sampling (CVS), the counselor was inundated with requests for CVS for sex selection. She refused, but occasionally a client would manipulate the system and make up an indication. The counselor was overworked, and eventually during a counseling session when it became clear that a prenatal couple was only interested in the fetal sex, she became quite angry. She cut short the session by truncating the information presented. She justified her actions by assuming that the couple wasn't interested in the information anyway.

- A genetic counselor with a history of infertility met with a client who went through infertility treatment, and the counselor asked the client several questions about procedures, feelings, and how the client got pregnant—all unrelated to the genetic question, but of interest to the genetic counselor.

- A genetic counselor who had a new baby met with a client who had a 6-month-old and was pregnant again. The genetic counselor commented that it must be a shock to be pregnant again, and that the client must be very tired. The client informed the counselor that this was a planned pregnancy and that she was not shocked.

- A genetic counselor who had an abortion as a teenager (and regrets that decision) viewed a client's indecision as wanting someone to tell her that it's OK to keep the baby and to raise it as a single mother. Without meaning to influence the client's decision, the counselor explained that the problems seen on the ultrasound were not all that bad, and that the baby would most likely still have a very good quality of life. The counselor just didn't want the client to "make the same mistake I did."

- A genetic counselor who had struggled with 18 months of active infertility treatment believed that her client's attitude toward pregnancy was all wrong. The client was young, naive, and afraid of needles, so she refused to have any of the available screening/testing options that might help identify if there was a genetic problem in the pregnancy (anomalies had been seen on the ultrasound). The genetic counselor tried to get through to the client's mother, since the client couldn't be reasoned with. In doing so, she neglected her client's autonomy and tried to influence the client's family in order to reach her own personal goal.

- The few times that clients have refused to have a student attend a counseling session, the genetic counselor suspects that a transference issue is present. For instance, the client wishes to exert control over the session, as a reaction to possibly feeling powerless in the exchange. The counselor's countertransference in these situations is sometimes to assess the clients as manipulative even though she has not explored their reasons for limiting student contact.

Maybe they have had a bad experience with students in the past, or are simply private people, or any variety of other explanations.

- The genetic counselor has a bit of a problem with authority, so counseling lawyers and judges is a challenge for her. She comes to these sessions with some uneasiness. She had one judge offer to examine her agency's consent form outside of the session, and he produced a three-page brief on the form's merits and limitations. This simply confirmed her uneasiness in these sessions.

- A genetic counselor found herself experiencing countertransference with clients who are statisticians. This group was particularly grueling for her to deal with when discussing MSAFP screening. The algorithm for value calculation is dissected, and the issue becomes the process, and not the implications of testing. It is as if there is no baby in the equation, just a risk assessment problem. The genetic counselor found herself dreading these sessions and knew that she could easily become angry with the clients for disregarding the baby.

Resolution of Countertransference Feelings

There are several strategies you can try in order to recognize and manage your countertransference:

- Accept the inevitability of countertransference. It happens to everyone sometimes. It does not mean that you are a bad genetic counselor or a bad person. An accepting, nondefensive attitude is essential.

- Locate the source of feelings. Ask yourself, "I wonder why this is so. Why did I make this particular response to this person's remark? What was behind it? What was I reacting to when making this remark? Why did I ask that question? Was it really to help my client?"

- Seek supervision assistance.

- Discuss your reactions with your client. (See Chapter 10 on self-reference behaviors.) Be careful not to shift the focus of the session to you, and do not disclose intense countertransference feelings.

- Analyze sessions with your supervisor (the use of audiotapes or videotapes can be very helpful).

- Engage in self-reflection and seek feedback. Since countertransference is mostly unconscious, it usually won't be detected until after it has happened (Cerney, 1985). This is why self-reflection and supervision are so crucial. You might realize after a genetic counseling session that you were acting out of character, or your supervisor or another colleague might comment on the intensity of your feelings about a client or situation (either intense like or dislike, intensely defensive, etc.). These are clues to countertransference. After a challenging genetic counseling session, it may be useful to ask yourself what, if anything, was out of character for you (Cerney, 1985).

- Seek personal counseling/psychotherapy if your countertransference is fairly pervasive and you are unable to manage it with the preceding strategies.

Stress and Burnout

Stress, Burnout, and Taking Care of Yourself

Genetic counseling requires intense involvement with clients on a highly personal level. Both technical expertise and sound emotional health are necessary to meet the demands of the profession. The practice of genetic counseling can be very stressful. Often clients have genetic conditions and situations for which there is no remedy. Can you imagine sitting hour after hour, day after day, listening to client stories and empathizing with them? In such a situation, you may find yourself beginning to experience your clients' intense emotions of anger, grief, despair, and fear, and you may have trouble letting go of these emotions. Additionally, whether you are at the beginning of your career or have been in the genetic counseling profession for a number of years, you face the continuing challenge of remaining involved and satisfied with your work. This can be difficult because stress and burnout are common in the helping professions (Fine and Glasser, 1996; Geldard, 1989; Skovholt, 2000).

What are stress and burnout? They are physical, emotional, and cognitive reactions to overload. There are a number of signs and symptoms, including feeling emotionally drained, overwhelmed, and out of control; physical, mental, and emotional fatigue; feeling reluctant or dreading going to work and seeing clients; physical reactions such as headaches, stomach complaints, and back pain; having a cynical attitude toward clients and/or feeling too detached from them; lack of satisfaction; and questioning whether you are being helpful or are doing anything meaningful (Geldard, 1989).

One factor that leads to stress and burnout is taking on too many commitments to the point where you feel as if you are always working and always behind (Volz, 2000). Another common risk factor is working alone or in isolation from others (Geldard, 1989; Volz, 2000). The less time you make for yourself and the more isolated you become in dealing with your reactions to your work, the more likely you are to develop stress and burnout symptoms (Fine and Glasser, 1996; Geldard, 1989; Volz, 2000). Another factor involves difficulties maintaining appropriate boundaries in your counseling work. Although you must develop an empathic connection with clients in order to be effective, you must be careful about becoming overly involved: "With experience you will learn how to walk beside the client with empathy and also how to protect yourself from the excesses of emotional pain by at times moving back a little, grounding yourself, and then joining with the client again" (Geldard, 1989, p. 177).

Since professionals can experience stress and burnout at any time in their career (Geldard, 1989), you must develop ways to alleviate the symptoms and to either reduce or remove the underlying causes. Geldard (1989) recommends several strategies:

- Recognize and acknowledge what you are experiencing (this strategy alone may reduce some symptoms).
- Talk with someone (a supervisor, colleague, trusted friend) about your feelings, so you can gain perspective.
- Reschedule your work to give yourself a feeling of control.
- Reduce your workload.
- Take a vacation.
- Practice relaxation or meditation exercises.
- Use positive self-talk.
- Lower your expectations of yourself, your clients, your peers, and your employer (you will get into difficulty if you have perfectionistic standards).
- Enjoy life, and cultivate and use your sense of humor.
- Do not take your problems home with you. Use thought-stopping techniques to stop worrying about clients when you are not at work. For instance, if you catch yourself worrying, try distracting yourself, or tell yourself that you will deal with the worry tomorrow at work at a specific time, and then focus on what you are doing in the here-and-now.
- Use a religious or philosophical belief system for support.
- Try to lead a balanced life that includes leisure and relationships, as well as work.
- Schedule the first 30 minutes of each day to collect your thoughts and prepare for the events ahead of you.
  Additional strategies include:
- Participate in peer supervision. Peer supervision can act as a buffer against stress and burnout, can help you manage your genetic counseling cases, and can strengthen your clinical skills (Kennedy, 2000).
- Seek personal counseling or psychotherapy to deal with crises you are experiencing at work (Volz, 2000).
- Make friends with people who are outside of the genetic counseling field (Volz, 2000).
- Give yourself some real downtime where you turn off the computer, the fax, and put away the paperwork (Volz, 2000).
- Schedule time during the day (mid-morning, mid-afternoon) that is just for you to take a walk, relax your muscles, meditate, etc. (Fine and Glasser, 1996).
- Think in sane ways (Fine and Glasser, 1996). To a great extent stress comes from what we tell ourselves about what we experience rather than from the experience itself. For example: "I have to be really helpful with every client I see"; "I can't make any mistakes or everyone will think I am incompetent";

"If I were a good counselor, I wouldn't be feeling so overwhelmed"; "If I were a good counselor, I would have all of the information in my head." Try to replace these irrational ideas with more reasonable ones.

Stress as a Novice

Although stress and burnout can occur at any time in your career, you may be experiencing unique stresses because you are learning how to become an effective helper. One study (Nutt Williams et al, 1997) analyzed the personal reactions and concerns of counseling psychology trainees during their prepracticum clinical work. Typical negative reactions in early counseling sessions included anxiety; discomfort; feeling distracted, unengaged, or self-focused; frustration/anger; and feeling inadequate or unsure of one's self. The trainees had four major concerns: (1) their therapeutic skills (e.g., performance anxiety); (2) their therapeutic role (defining the role and its boundaries; for example, some trainees wanted to jump in and solve client problems, some felt responsible for client feelings, etc.); (3) difficult clients and conflict in the session (e.g., client resistance, client distance); and (4) reactions to specific content (e.g., feeling sad about a client's poor relationship with a family member).

How can you deal with the stresses of being a novice? The following strategies can be effective:

- Discuss how things are going with your peers and with friends.
- Talk with supervisors.
- Focus on positive feedback about your work and don't dwell solely on negative feedback.
- Engage in positive self-talk (e.g., "Here is what I did well. I made this mistake, but it's a common mistake for beginners").
- Maintain a sense of humor.
- Try to focus more on the client and less on yourself during sessions (Nutt Williams et al, 1997).
- Try trusting your own feelings and instincts about the right way to respond to clients (Nutt Williams et al, 1997).
- Observe others providing genetic counseling so that you have multiple models from which to develop your own style (Hendrickson et al, 2001).
- Take risks: volunteer to do role-plays, disclose your concerns in supervision and in classroom discussions, etc.
- Keep a journal for recording your thoughts and feelings. Periodically review earlier entries so that you can see how you are developing as a genetic counselor.
- Practice in simulated genetic counseling sessions with a classmate or friend and record your sessions. Play them back and critique your work (what went well and not so well).

- Seek personal counseling if your anxiety is having a negative impact on your clinical work and/or your responsiveness to supervision.
- Set your genetic counseling priorities. Decide what are the most important things to accomplish during a session and what things are less important.

Impact of Personal Counseling on Genetic Counseling Practice

Personal counseling can be an effective way to deal with general stress and burnout and to cope with the unique stressors of being a novice. In addition to helping you resolve these situations, personal counseling or psychotherapy can have other positive effects. Some research (e.g., Macran and Shapiro, 1998; Macran et al, 1999; Peebles, 1980) suggests that personal therapy has one or more of these benefits for clients who are themselves counselors/therapists:

- Leads to greater empathy, acceptance, and genuineness, possibly because therapy increases one's ability to be aware of feelings and makes one more comfortable discussing affect.
- Allows a therapist to know how it feels to be a client.
- Provides a role model of how to counsel on psychosocial issues (e.g., using gestures, expressions, and techniques that their therapists used with them).
- Allows therapists to learn to be themselves.
- Helps therapists learn their own limits and boundaries.
- Helps therapists learn what not to do (i.e., therapist behaviors that were not helpful for them).
- Helps therapists learn how to separate their own feelings from their clients' feelings.
- Helps therapists address issues on a deeper level with their clients.

Although these studies are based on mental health counselors, to the extent that genetic counseling shares some common elements, we would argue that personal counseling or psychotherapy might be beneficial for genetic counselors.

Closing Comments

For all of the issues we discussed in this chapter, it is important to remember that they are common occurrences. You are not the only one to have a countertransference reaction or to feel stress and burnout; and you are not weird or incompetent because of these experiences. The worst thing that you can do is pretend that these things are not happening to you. Unacknowledged, they will only get worse. The best thing you can do is to be proactive—acknowledge what is going on and deal with it by consulting with a supervisor, a professor, a genetic counselor, and peers.

Class Activities

Activity 1: Transference Discussion:
Think-Pair- Share Dyads

Students discuss the following situation: Your client expresses unrealistic expectations about what you should be able to do for her. She tells you that you should be able to calculate an absolute risk rate for her pregnancy, that you should be able to do genetic testing to guarantee that her baby will be all right, and that if you cared more about her situation you would tell her what she should do if abnormal test results come back.*

- How would you deal with her unrealistic expectations?
- Would you try convincing her that her demands are unrealistic? How would you do this?
- Would you be inclined to give her advice by telling her what you think she should do if she receives abnormal test results?

PROCESS

The whole group discusses their responses to the situation.

Estimated time: 30 to 45 minutes.

Activity 2: Transference
ROLE-PLAY

Using the situation in Activity 1, engage in a 5- to 10-minute role-play in which the genetic counselor attempts to respond to client expectations. This can be done in triads or in small groups.

PROCESS

Discuss in triads or in the whole group the counselor's reactions and responses to the client, the client's reactions to counselor interventions, and how transference can be managed in genetic counseling sessions.

Estimated time: 60 to 75 minutes.

Activity 3: Countertransference Discussion:
Think-Pair-Share Dyads

Students discuss with a partner the questions in written Exercise 2. They can do this either before or after completing the exercise.

Estimated time: 20 to 30 minutes.

---

*Adapted from Corey et al, 1984.

Activity 4: Countertransference

ROLE-PLAY

In triads, the student who will be the counselor selects one of Watkins' four types of countertransference and plays this out in a 10- to 15-minute role-play. The student should not tell the other members of the triad which type of countertransference will be demonstrated, and s/he should be subtle in showing the countertransference.

PROCESS

Triads discuss the type of countertransference they thought the counselor demonstrated and the counselor behaviors that indicated the countertransference. Next, they should discuss the client's reaction to the counselor's behaviors. They can also discuss what the counselor could have said or done differently to manage the countertransference and counsel more effectively.

Estimated time: 75 to 90 minutes.

INSTRUCTOR NOTE

- The instructor could conduct a large group discussion of student reactions to the role-plays on transference and countertransference as well as their thoughts about these two counselor-client dynamics.

ACTIVITY 5: STRESS AND BURNOUT DISCUSSION: THINK-PAIR-SHARE DYADS

Students take turns talking about a situation from their past where they experienced stress and burnout and what they did to cope with the experience. Next, they discuss what aspects of a career in genetic counseling might make them vulnerable to stress and burnout.

PROCESS

The whole group offers strategies for dealing with stress and burnout. The group can also discuss what aspects of genetic counseling may make them personally vulnerable to stress and burnout.

Estimated time: 45 minutes.

INSTRUCTOR NOTE

- This activity could also be turned into a written exercise—either as a follow-up to the activity or instead of the activity.

Written Exercises

Exercise 1: Irrational Beliefs

List all of the irrational beliefs that you have about being a genetic counselor, a supervisee, and a professional in the genetic counseling field. Then dispute each belief by writing down a more reasonable idea.

EXAMPLE

Irrational belief: "I must be really helpful with all of my clients or I am not a good genetic counselor." Disputing belief: "I should try to be helpful to my clients while recognizing that some will find what I have to say more helpful than will others."

Exercise 2: Countertransference Issues

Describe the types of clients (e.g., terminally ill, highly emotional, etc.) and genetic counseling situations (e.g., sex selection, clients who refuse to reveal information to at-risk relatives) that "push your buttons" or that you find particularly hard to work with as a genetic counselor. Discuss what makes these clients/situations so challenging, and then describe the type of countertransference you would be likely to use in response to these clients and situations.

Exercise 3: Countertransference Checklist*

Place a check in the space to indicate whether you have had this tendency or believe you might be likely to have this tendency in future genetic counseling work:

__I am likely to become overly involved on an emotional level with certain clients.
Comment:

__I might find a number of ways to avoid clients who prompt strong negative feelings in me.
Comment:

__I worry that I may have a strong need to give lots of advice, and that I will manipulate clients to think and act the way I think they should.
Comment:

__I might fall back on giving excessive amounts of information to clients in order to keep the session structured and emotionally safe.
Comment:

__I am concerned that I may bring home some clients problems, and I will overidentify with some of my clients.
Comment:

---

*Adapted from Corey et al, 1984.

__I can imagine myself getting angry and upset over clients who do not appreciate me.
Comment:

__I tend to respond very defensively to certain types of people or certain kinds of remarks.
Comment:

__There are some topics I would feel very uncomfortable exploring with clients, and I am likely to steer away from talking about them.
Comment:

__I am afraid that I will feel responsible if a client does not understand, cannot make a decision, or makes a bad decision.
Comment:

__I worry that I will pity my clients who have disabling or terminal conditions.
Comment:

__I am afraid that I would break down and cry with some clients.
Comment:

__I usually do whatever I can to avoid negative or angry encounters.
Comment:

Annotated Bibliography

Ball, S. (1988). Coping with burnout in genetic counseling. In: Ball, S., ed. Strategies in genetic counseling: the challenge of the future, vol. 1. National Society of Genetic Counselors Series. New York: Human Sciences Press, 109-116.
[Describes a burnout model and discusses stresses in genetic counseling and counselor personal reactions.]
Dalenberg, C. (2000). Countertransference and the treatment of trauma. Washington, DC: American Psychological Association.
[Provides useful strategies for responding to client trauma and offers examples of countertransference experiences from experienced psychotherapists.]
Djurdjinovic, L. (1998). Psychosocial counseling. In: Baker, D. L., Schuette, J. L., Uhlmann, W. R., eds. A guide to genetic counseling. New York: John Wiley and Sons.
[Discusses transference and countertransference as examples of disruptions in the genetic counseling relationship.]
Guy, J. (1987). The personal life of the psychotherapist. New York: John Wiley and Sons.
[Provides a personalized account of the reciprocal impact of one's life experiences and one's experiences as a psychotherapy practitioner.]
Journal of Genetic Counseling (2000) Special issue: supervision for practicing genetic counselors. 9: 375-434.
[Discusses the role of supervision for genetic counselors. Members of a leader-led supervision group describe its impact on their professional development and practice.]
Kessler, S. (1992). Psychological aspects of genetic counseling. VIII. Suffering and countertransference. Journal of Genetic Counseling, 1, 303-308.

[Discusses countertransference within genetic counseling and gives examples of client issues that may trigger it.]

Skovholt, T. S. (2000). The resilient practitioner: burnout prevention and self-care strategies for counselors, therapists, teachers and health professionals. Boston: Allyn and Bacon.
[Describes sources and symptoms of burnout and suggests techniques for addressing burnout.]

Watkins, C. E., Jr. (1983). Transference phenomena in the counseling situation. Personnel and Guidance Journal, 64, 206-210.
[Defines client transference and gives specific examples.]

Watkins, C. E., Jr. (1985). Countertransference: its impact on the counseling situation. Journal of Counseling and Development, 63, 356-359.
[Proposes four major types of counselor countertransference and discusses implications for clinical work.]

Weil, J., and Mittman, I. (1993). A teaching framework for cross-cultural genetic counseling. Journal of Genetic Counseling, 2, 159-169.
[Discusses the importance of genetic counselor cultural self-awareness to prevent cultural countertransference and recommends ways that training programs can increase student cultural knowledge and sensitivity.]

CHAPTER 13

Using Internet Resources

Learning Objectives

1. Identify appropriate Internet resources for providing patient care.
2. Recognize limitations that are inherent in these resources.

Use of Internet Resources

Developing strategies for keeping up to date with new information in order to appropriately serve patients with new diagnostic and testing options is a major challenge for genetic counselors. This challenge will intensify as the amount and complexity of the information grows and as specialized areas of practice develop in the profession. Since competency is a core component of providing safe patient care, genetic counseling practitioners need to develop a bank of resources that they can refer to for the most up-to-date information. The Internet is one valuable source of such information. In addition to assisting in the search for medical genetic information, the Internet serves as a source of support information for families dealing with a genetic disease, is a source of referral information for other health care practitioners and patients and families, and provides a way of connecting with colleagues in the genetics profession and in related professions.

A number of Web sites make genetic information available so practitioners can explore the nature of specific conditions. There are excellent links to information about genetic conditions, research, and genetic tests and they are updated on a regular basis. Professional societies maintain Web sites that serve as helpful resources for health care professionals for referral information and consultation. Additionally, there are excellent patient and family advocacy groups with information available at the lay level and sometimes in multiple languages. Patients, families, and other health care providers access these sites along with the genetics professionals.

Remember, as a genetics health care provider you are a resource to colleagues, patients, families, and other health care professionals:

- Keep your contact information listed in professional directories such as the National Society of Genetic Counselors (NSGC) and American Society of Human Genetics (ASHG) Web sites up to date. Patients, families, and other health care professionals use these resources to find you when they are searching on the Web.

- Consider developing your own Web site or a site for the genetics services at your center. Such a site could describe your services so that patients, families, and other health care professionals in your area can find you. In addition, this site could fill an educational need with information about developments in medical genetics.

- Participate in relevant professional listservs. This can help you obtain critical information in a rapid fashion. However, it is a privilege to be a participant of a listserv, so remember to give as well as take from this resource. If you participate in a listserv, do your homework before asking for help with a case. Do not expect others to look up basic information for you. Share what you have learned about the condition/case and let others know the source of your information.

- Remember that when you put something in writing, the conditions of the communication are different than in person. People cannot see your face and hear your voice. These are cues that people use to decipher the actual meaning of your communication such as your intent, confidence, level of support, etc. Write carefully and read what you have written before you send it out. Think twice about sending an inflammatory message or response. On-line messages are not private and many people may see your message.

- Always protect the privacy of the patient. Remember when presenting a case for consideration by a colleague that identifying information about a patient or family should never be included. E-mails are a written communication medium. They can be retrieved long after they are sent. Write with care.

- Use the same caution in responding to e-mail messages from patients requesting information that you use when discussing issues over the phone. Remember that genetic counseling is not useful, and actually can be very harmful, if the information given pertains to the incorrect diagnosis. Since it is not possible to verify a diagnosis over the Internet, take care when giving out critical medical information that may be used by a patient to make important life decisions.

- Use your hospital/institutional attorney as a resource. If you have any questions about a message you received or whether or not, or how, to answer a message, check with your attorney.

Limitations Involved with Internet Resources

Although the Internet is a valuable source of information, it is also a source of much misinformation. You should draw upon the same skills you use to evaluate published work when you search the Internet for information. Be cautious about relying on information from the Internet; its easy accessibility makes it a tempting source for all of your information and therefore it may be equally as easy to overlook poor quality and misinformation. Gaining a full understanding of the genetic and medical aspects as well as the psychosocial ramifications of a disorder and the available testing options takes much time and effort. Short cuts rarely pay off.

When using the Internet as a resource, be sure to consider the following:

- Consider the source of the information on the Web site. Is it credible? Data from a paper published in a respected peer-reviewed journal are more credible than data reported in a popular magazine, for example. In the same way, information on a Web site from a group or center that includes respected professionals in the field is more credible than information on a popular public Web site.

- Look at the authors of the Web site. Always look for a listing of the professionals involved with the Web site. Call or e-mail with questions you have, and be concerned if there is no way to easily contact the person with whom you would like to communicate.

- When searching for technical information, look for corroborating data. Is the same information available from more than one source? Is it published anywhere and if so, where? If more than one credible source is reporting the same information, it is more likely to be good information.

- Watch for conflicting information. If you find conflicting data, check it out with data published in a peer-reviewed article or a recent reference text.

- Contact experts in the field when the information you need is not clear. Often there are a few researchers (or sometimes only one researcher) who are considered to be experts with specific conditions. Locate these persons by checking out who has published most of the available information. Contact these individuals directly for the best information. Second-hand information is often lacking in accuracy.

- When using the Internet to explore testing options, contact more than one laboratory. If only one lab offers the test you are interested in, why is that? Is it because no other lab can duplicate the test? How reliable is the test if this is the case? Is this a clinical test or a research test? Are there experts available to talk with you?

- Expect patients and families increasingly to come in with information they have obtained from an Internet source. In many cases, this information will contain inaccuracies. Approach this situation as you would those in which patients and families come in with other sources of misinformation. Your strategy should include having information available that you know is credible and discussing how you evaluate available information.

- Never refer a patient, family, or colleague to an Internet source that you have not evaluated. There is too much misinformation out there to assume that every resource is a good one.

Examples of Internet Resources for Genetic Counseling

The following is a collection of Web sites that are commonly used in the practice of genetic counseling. Remember that Web sites, like other sources of published information, become dated quickly if not maintained and frequently updated; new sources develop continually. Use these resources wisely and remember that your best source of information is based on data that have been published recently in a credible peer-reviewed journal.

American Academy of Pediatrics (AAP)

Web address: http://www.aap.org/

Maintained by: American Academy of Pediatrics

Audience: mostly for pediatricians but also much information for the lay public.

Restrictions: mostly accessible but has a member's only link.

Mission statement: "The mission of the American Academy of Pediatrics is to attain optimal physical, mental and social health and well-being for all infants, children, adolescents and young adults. To this purpose, the AAP and its members dedicate their efforts and resources."

Contents: The site carries many news releases and maintains a special page listing Internet resources for pediatricians, including a special listing of genetics Web sites. This is located under the professional education section under the Web address: http://www.aap.org/bpi/Genetics.html.

American College of Medical Genetics (ACMG)

Web address: http://www.faseb.org/genetics/acmg/

Maintained by: American College of Medical Genetics

Audience: clinical genetics specialists

Restrictions: mostly accessible but has a member's only link.

Mission Statement: "The ACMG provides education, resources and a voice for the medical genetics profession. To make genetic services available to and improve the health of the public, the ACMG promotes the development and implementation of methods to diagnose, treat, and prevent genetic disease."

Contents: Several in-depth modules describing current standards of practice, knowledge, testing protocols, and policy recommendations regarding genetic testing. They are not interactive, but fairly in-depth. This site is useful for genetics health professionals or for health professionals with a basic understanding of genetics.

American College of Obstetrics and Gynecology (ACOG)

Web address: http://www.acog.org/

Maintained by: American College of Obstetrics and Gynecology

Audience: OB/GYNs, other health care professionals, and the general public.

Restrictions: mostly accessible but has a member's only link.

Mission statement: "The American College of Obstetricians and Gynecologists (ACOG) is a membership organization of obstetrician-gynecologists dedicated to the advancement of women's health through education, advocacy, practice, and research."

Contents: Mostly provides information for obstetricians and gynecologists about the organization, meetings, etc. However, there is quite a bit of information for the lay public about pregnancy issues and the risk for birth defects. Also, this site publishes ACOG position statements on issues such as population screening for cystic fibrosis.

American Medical Association Genetics and Molecular Medicine

Web address: http://www.ama-assn.org/ama/pub/category/1799.html.

Maintained by: American Medical Association

Audience: primarily physicians, but also other health care professionals and patients.

Restrictions: mostly accessible but has a member's only link.

Mission statement: "To promote the science and art of medicine and the betterment of public health. Founded more than 150 years ago, AMA's strategic agenda remains rooted in our historic commitment to standards, ethics, excellence in medical education, practice, and advocacy on behalf of the medical profession and the patients it serves."

Contents: Provides an educational primer on basic genetics, policy statements, research updates, news announcements, and an on-line continuing medical education (CME) module exploring some of the issues in breast cancer testing.

American Society of Human Genetics (ASHG)

Web address: http://www.ashg.org/genetics/ashg/ashgmenu.htm.

Maintained by: American Society of Human Genetics

Audience: members of the society.

Restrictions: none.

Mission statement: "ASHG serves research scientists, health professionals, and the public by providing forums to:

- Share research results at annual meetings annual and in the American Journal of Human Genetics.
- Advance genetic research by advocating for research support.
- Enhance genetics education by preparing future professionals and informing the public.
- Promote genetic services and support responsible social and scientific policies."

The American Society of Human Genetics (ASHG), founded in 1948, is the primary professional membership organization for human geneticists in North America. Over 5,000 members include researchers, academicians, clinicians, laboratory practice professionals, genetic counselors, nurses, and others involved in human genetics. The principal objectives of ASHG are to bring together investigators in the many areas of endeavor that involve human genetics, and to encourage and integrate their efforts by providing a forum for sharing research findings. This is accomplished primarily through the society's annual meeting and its official monthly publication, the American Journal of Human Genetics.

Contents: Information on careers in genetics, public policy statements on current issues, a searchable membership directory, and information that is important to members. This site can connect the user to a medical geneticist, research geneticist, or genetic counselor.

Foundation for Genetic Education and Counseling

Web address: http://www.fgec.org/

Maintained by: Foundation for Genetic Education and Counseling

Audience: general public and all health care professionals (nongenetic).

estrictions: none.

Mission statement: "The Foundation for Genetic Education and Counseling will promote genetic education for the general public and for health-care professionals. The foundation will develop, evaluate, and disseminate educational programs for individuals affected by or at risk for common, complex diseases and will provide information and resources that inform genetic counseling."

Contents: Two noninteractive modules. One is an in-depth explanation of the genetics of complex diseases. The other is a case study of genetics and bipolar disease. Explanations of genetic counseling are included. Excellent educational material is available on this site.

Genetic Alliance

Web address: http://www.geneticalliance.org/

Maintained by: The Genetic Alliance

Audience: anyone interested in lay information about genetic disorders.

Restrictions: none.

Mission statement: "The Genetic Alliance is an international coalition of individuals, professionals and genetic support organizations that are working together to promote healthy lives for everyone impacted by genetics."

Contents: Policy statements, initiatives, publications, connections to many advocacy and support groups, and patient literature. This is an extremely valuable resource for genetic counselors and anyone working with families who have a genetic condition.

The Genetics of Cancer

Web address: http://www.cancergenetics.org/

Maintained by: Northwestern Medical School

Audience: mostly the general public, but has a link for health care professionals.

Restrictions: none.

Mission Statement: "The Genetics of Cancer is a resource center designed to help you understand the genetic basis of cancer and interpret new discoveries in the field of cancer genetics. Here you will find both clinical and basic information on cancer, heredity, and the roles that genes can play in the development of various cancers. You can search for a specific topic, step page by page through the material by clicking the "next" button at the bottom of the screen, or just see where the hyperlinks—and your curiosity—take you."

Contents: A very detailed overview of issues in genetic testing for cancer. Provides three different levels of explanation. Discusses a range of issues including how to ask patient questions, how to take a family history, and how to provide genetic counseling. Provides 11 different case studies to illustrate the issues. Has links on how to obtain additional resources.

Genetic Health

Web address: http://www.genetichealth.org/

Maintained by: private company founded by former Celera Genomics employee and applied biosciences employee.

Audience: general public.

Restrictions: none (there is an option to become a member in order to receive personalized updates on genetics—based on one's risks as determined by this Web site.)

Mission statement: "Our mission at Genetic Health is to provide Internet-based products and services that enable health companies, research organizations, and health care professionals to move genetic research advances out of the lab and into patients' lives."

Contents: Essentially a tool for the general public to understand genetic risks—and to assess their own risks for conditions such as cancer, Alzheimer's, and other common conditions. Tools for the general public to create their own family pedigree—promises to have a new feature in which an individualized cancer risk can be calculated by the Web site. Individuals can become members and receive updates on conditions for which they are at risk.

Genetics Education Center

Web address: http://www.kumc.edu/gec/

Maintained by: University of Kansas

Audience: educators (high school), general public.

Restrictions: none.

Mission statement: "For educators interested in human genetics and the human genome project."

Contents: Links to on-line lesson plans that cover every imaginable aspect of genetics, genetic testing, human genome project. Lesson plans cover technical aspects, legal aspects, and social aspects of genetics. The site itself doesn't have a lot of information but provides numerous of links to lesson plans. Additionally, the site has links to a multitude of patient advocacy groups in the United States and around the world. This is an excellent resource for practicing genetic counselors seeking patient advocacy referral information and lay literature and anyone who is interested in teaching basic genetics.

Genetics Education and Counseling Program (GECP)

Web address: http://www.pitt.edu/~edugene/

Maintained by: University of Pittsburgh (funding from NCI, Genzyme, others)

Audience: general public, primary care practitioners.

Restrictions: none.

Mission statement: "The Genetics Education and Counseling Program, a joint effort of the University of Pittsburgh and UPMC Health System, is dedicated to providing up-to-date information about inherited conditions and related services for individuals, families, and whole communities."

The program is designed to: Translate advances in genetics into public health services, education, awareness, and policy. Improve awareness of genetics and genetic diseases in the community and among professionals. Provide information about individual referrals to clinical and testing services in genetics. Offer important information on diagnosis, treatment, and resources available to people to cope with the medical, financial, emotional, and social burdens of the illnesses for which they are at risk. Provide a guide to services for affected individuals and families. Safeguard genetic privacy. Serve as a guide and model for the development of other public health programs.

Program services include: Educational workshops and programs tailored to specific interests and needs of organizations and individuals interested in genetics, genetic diseases, and the range of genetics services available in Pittsburgh and western Pennsylvania (and elsewhere). Referral information for individuals and families that seek genetic evaluation and consultation, as well as genetic testing, when appropriate.

Resource/informational materials

Brochures and fact sheets

World Wide Web sites

Telephone hotline numbers

Comprehensive interactive on-line education modules about many hereditary and family-related conditions, including family-related cancers; susceptibilities to hereditary breast and/or ovarian cancer; and Tay-Sachs, Gaucher, and Canavan diseases.

"People who may benefit from the program include anyone seeking information about a particular genetic or family-based condition, genetics services, genetic testing, treatment, or genetics in general health care professionals who want to keep abreast of new developments in medical genetics as well as ethical and social issues that accompany new discoveries."

Contents: Provides numerous links to genetics education Web sites, and many brochures and fact sheets on genetic diseases. Brochures are aimed at the general public.

Gene Tests/GeneClinics

Web address: http://www.geneclinics.org/

Maintained by: University of Washington (grant from National Institutes of Health)

Audience: genetics professionals, primary care providers seeking genetic information.

Restrictions: open to all but a free of charge one-time registry process is required.

Mission statement: "A clinical information resource relating genetic testing to the diagnosis, management, and genetic counseling of individuals and families with specific inherited disorders."

Contents: Well-organized summaries of numerous genetic conditions. Information includes diagnosis, prognosis, genetic testing, and genetic counseling. Includes links to professional societies (ASHG, NSGC) and links to patient resources.

The National Cancer Institute (NCI)

Web address: http://www.cancer.gov/.

Maintained by: National Cancer Institute

Audience: anyone interested in information about cancers, research protocols, funding opportunities, and clinical trials.

Restrictions: none.

Mission statement: "The National Cancer Institute coordinates the National Cancer Program, which conducts and supports research, training, health information dissemination, and other programs with respect to the cause, diagnosis, prevention, and treatment of cancer, rehabilitation from cancer, and the continuing care of cancer patients and the families of cancer patients."

Contents: A description of the research programs administered by the NCI, research funding opportunities, information on all types of cancer and the clinical trials available for people with different types of cancer.

The National Center Birth Defects and Developmental Disabilities

Web address: http://www.cdc.gov/ncbddd/default.htm.

Maintained by: Center for Disease Control (CDC)

Audience: anyone interested in information about cancers, research protocols, funding opportunities, and clinical trials.

Restrictions: none.

Mission statement: "The National Center on Birth Defects and Developmental Disabilities (NCBDDD) at the Centers for Disease Control and Prevention (CDC) seeks to promote optimal fetal, infant, and child development; prevent birth defects and childhood developmental disabilities; and enhance the quality of life and prevent secondary conditions among children, adolescents, and adults who are living with a disability. Below are just some of the topics that our Center is currently working with in the areas of birth defects, developmental disabilities, and disabilities and health."

Contents: Provides information about childhood illnesses, birth defects, and developmental disabilities for the lay public. The site has a link for youth and for health professionals.

National Coalition for Health Professional Education in Genetics (NCHPEG)

Web address: http://www.nchpeg.org/

Maintained by: NCHPEG

Audience: health professionals—all types.

Restrictions: none.

Mission statement: "Catalyzed in 1996 by the American Medical Association, the American Nurses Association, and the National Human Genome Research Institute, the mission of the National Coalition for Health Professional Education in Genetics (NCHPEG) is to promote health professional education and access to information about advances in human genetics to improve the health care of the nation."

Contents: Essentially a multidisciplinary collaboration designed to promote genetics education in all health fields. Defines basic competencies for all health professionals, and provides education resources. The education resources are primarily links to other Web sites that contain information, case studies, tools (pedigree builders) etc.

The National Human Genome Research Institute (NHGRI)

Web address: http://www.nhgri.nih.gov/.

Maintained by: National Institutes of Health (NIH)

Audience: anyone interested in the human genome project.

Restrictions: none.

Mission statement: "The National Human Genome Research Institute (NHGRI) was originally established in 1989 as the National Center for Human Genome Research (NCHGR). Its mission is to head the Human Genome Project for the NIH."

Contents: Provides information on the Human Genome Project, grant opportunities, research programs, policy and public affairs, resources, and workshops. This site also has links to legislation concerning genetics issues such as protection of privacy and genetic discrimination.

National Society of Genetic Counselors (NSGC)

Web address: http://www.nsgc.org/.

Maintained by: National Society of Genetic Counselors

Audience: genetic counselors, people interested in genetic counseling as a career, people seeking genetic counseling, professional connections.

Restrictions: mostly accessible but has a member's only link.

Mission statement: "The NSGC is the leading voice, authority and advocate for the genetic counseling profession. The society promotes the genetic counseling profession as a recognized and integral part of health care delivery, education, research, and public policy."

Contents: Information for genetic counselors, and people interested in genetic counseling, and people seeking genetic counseling. News updates in genetics and a section called "Genetic Counseling and You" provides phone numbers and e-mail addresses of genetic counselors, so individuals can connect with a professional in their area.

Online Mendelian Inheritance in Man (OMIM)

Web address: http://www3.ncbi.nlm.nih.gov/Omim/.

Maintained by: Johns Hopkins University

Audience: primarily Genetics professionals.

Restrictions: none.

Mission statement: "Welcome to OMIM(TM), Online Mendelian Inheritance in Man. This database is a catalog of human genes and genetic disorders authored and edited by Dr. Victor A. McKusick and his colleagues at Johns Hopkins and elsewhere, and developed for the World Wide Web by the National Center for Biotechnology Information (http://www.ncbi.nlm.nih.gov)."

Contents: The database contains textual information, pictures, and reference information. It also contains copious links to NCBI's database of MEDLINE articles and sequence information. This is the most up-to-date information on known genetic diseases. This site is extremely valuable for genetic counselors and all health care providers.

Closing Comments

The Internet is powerful tool. It can assist the genetic counselor in her never-ending quest to keep current with the advances in the field. It is also a valuable means of maintaining essential connections with colleagues. However, it is critical to remember the limitations of this resource and to use it wisely.

Summary points to consider:

- Establish an ongoing relationship with colleagues. Remember, genetics is a team sport. It is virtually impossible to keep up with everything in genetic medicine on your own.

- Be critical of consumer Internet resources. Use Web sites such as those provided in this chapter, but be cognizant of their limitations.

- Remember to check out any conflicting information. Compare this information to published data.

- Learn about telemedicine resources available to you. This technology enables specialists to offer services and consultations to patients and families as well as other health care providers in remote areas of the country and connects colleagues.

- Familiarize yourself with the published policies of professional organizations. Most professional organizations have published policies addressing complex issues such as genetic testing in children, population screening, and others. Know those policies and call upon them when you are dealing with difficult cases.

- Keep reference texts up to date. In the arena of genetic medicine, texts are often outdated shortly after they are published.

Class Activities

Activity 1: Small Group Role-Plays

Break up into groups of two or three students. Each student takes a turn playing a genetic counselor and client (and a partner, mother, sibling, etc.). The client comes into clinic with information obtained from an on-line source that contains much misinformation. The counselor must figure out a way to give the client appropriate information and explain why the appropriate information conflicts with the information they obtained from the Internet. Each role-play should last for about 15 to 20 minutes.

PREPARATION

Students may need to gather appropriate information before class on the specific condition to be discussed in the role-play.

PROCESS

The students discuss various strategies of dealing with this difficult situation.

Estimated time: 90 minutes.

Activity 2: Dyads and/or Small Group

Students brainstorm strategies for answering e-mail requests for medical and genetic information from patients. Have students think of a request for information from a patient and then have the groups write an appropriate e-mail response.

PROCESS

The whole group discusses the responses that they generated.

Estimated time: 45 minutes.

Written Exercises

Exercise 1

Gather medical, genetic, research and lay information from various Web sites on the conditions listed below. Summarize the information available from each site and the unique contributions of the different sites.

Condition	Source of Information
• Down syndrome	
• Phenylketonuria (PKU)	
• Neurofibromatosis	
• Marfan syndrome	
• Cystic fibrosis	
• Hemophilia	
• Breast cancer	
• Familial adenomatous polyposis	

Exercise 2

This is a variation on Exercise 1. Compare the genetic and medical information you are able to obtain from on-line resources on the above-mentioned disorders with information in recently published peer reviewed journals. How does it compare? What are the differences, if any? Similarities?

Annotated Bibliography

American Medical Association (posted 3-17-2000). Guidelines for medical and health information sites on the Internet. http://www.ama-assn.org/ama/pub/category/1905.html.
[Discusses AMA guidelines for Web sites including principles for content, advertising and sponsorship, Web site privacy and confidentiality, and E-commerce.]
National Alliance of Breast Cancer Organizations. (2002). Evaluating Medical Information about Breast Cancer. http://www.nabco.org/index.php/8/index.php/149.
[Provides key questions to ask about a medical site.]
Tillman, H. N. (May 30, 2000). Evaluating quality on the Net. http://www.hopetillman.com/findqual.html
[This piece was originally created in 1995 with the title, "Evaluating the Quality of Information on the Internet or Finding a Needle in a Haystack" as a presentation delivered at the John F. Kennedy School of Government, Harvard University, Cambridge, Massachusetts, September 6, 1995.]
The World Wide Web Virtual Library. (October 2, 2002). http://www.vuw.ac.nz/~agsmith/evaln/evaln.htm
["This page contains pointers to criteria for evaluating information resources, particularly those on the Internet. It is intended to be particularly useful to librarians and others who are selecting sites to include in an information resource guide, or informing users as to the qualities they should use in evaluating Internet information."]
Wyatt, J. C. (1997). Commentary: measuring quality and impact of the World Wide Web. British Medical Journal, 314, 1879-1881.

References

Allport, F. H. (1924). Social psychology. Boston: Houghton Mifflin.

American Society of Human Genetics. (1975). Genetic Counseling. American Journal of Human Genetics, 27, 240-242.

American Society of Human Genetics Social Issues Subcommittee on Familial Disclosure (1998). Professional disclosure of familial genetics information. American Journal of Human Genetics, 62, 474-498.

Azar, B. (1997, November). Defining the trait that makes us human. American Psychological Association Monitor, 28, 1 & 15.

Baker, D. L. (1998). Interviewing techniques. In: Baker, D. L., Schuette, J. L., Uhlmann, W. R., eds. A guide to genetic counseling. New York: John Wiley & Sons, 55-74.

Barkham, M. (1988). Empathy in counselling and psychotherapy: present status and future directions. Counselling Psychology Quarterly, 1, 407-428.

Barrett-Lennard, G. T. (1981). The empathy cycle: refinement of a nuclear concept. Journal of Counseling Psychology, 28, 91-100.

Bartels, D. M., LeRoy, B. S., McCarthy, P., Caplan, A. L. (1997). Nondirectiveness in genetic counseling: a survey of practitioners. American Journal of Medical Genetics, 72, 172-179.

Bartels, D. M., McCarthy Veach, P., LeRoy, B. S. (2001). Genetics in primary care: implications for physician education. Unpublished manuscript. University of Minnesota.

Beauchamp, T. L., Childress, J. L. (1994). Principles of biomedical ethics (4th ed.). London: Oxford Press.

Bell, N. K. (1990). Medical ethicist responds to issue of nondirectiveness in genetic counseling setting. Perspectives in Genetic Counseling, 12, 5.

Bellet, P. S., Maloney, M. J. (1991). The importance of empathy as an interviewing skill. Journal of the American Medical Association, 266, 1831-1832.

Benkendorf, J. L., Callanan, N. P., Grobstein, R., Schmerler, S., Fitzgerald, K. (1992). An explication of the National Society of Genetic Counselors (NSGC) Code of Ethics. Journal of Genetic Counseling, 1(1), 31-38.

Bennett, R. L. (1999). The practical guide to the genetic family history. New York: Wiley-Liss.

Bertakis, K. D., Roter, D., Putnam, S. M. (1991). The relationship of physician medical interview style to patient satisfaction. Journal of Family Practice, 32, 175-181.

Beyene, Y. (1992). Medical disclosure and refugees—telling bad news to Ethiopian patients. In: Cross-cultural medicine a decade later [special issue]. Western Journal of Medicine, 157, 328-332.

Billings, P. R., Kohn, M. A., de Cuevas, M., Beckwith, J., Alper, J. S., Natowicz, M. (1992).

Discrimination as a consequence of genetic testing. American Journal of Human Genetics, 50, 476-482.

Bloom, B. S., ed. (1956). Taxonomy of educational objectives, handbook I: cognitive domain. New York: Longmans, Green.

Bottorff, J. L., Ratner, P. A., Johnson, J. L., Lovato, C. Y., Joab, S. A. (1998). Communicating cancer risk information: the challenges of uncertainty. Patient Education and Counseling, 33, 67-81.

Bower, J. A., McCarthy Veach, P., LeRoy, B. S., Bartels, D. M. (2002). Ethical and professional challenges: A survey of counselors' experiences. Journal of Genetic Counseling, 11, 163-186.

Brady, J. L., Guy, J. D., Poelstra, P. L., Brown, C. K. (1996). Difficult good-byes: A national survey of therapists' hindrances to successful terminations. Psychotherapy in Private Practice, 14, 65-76.

Brammer, L. M., Shostrom, E. L. (1982). Therapeutic psychology (4th ed.). Englewood Cliffs, NJ: Prentice-Hall.

Brown, D. (1997). Implications of cultural values for cross-cultural consultation with families. Journal of Counseling and Development, 76, 29-35.

Bryan, J. H. (1972). Why children help: a review. Journal of Social Issues, 28, 87-104.

Carlos Poston, W. S., Craine, M., Atkinson, D. R. (1991). Counselor dissimilarity confrontation, client cultural mistrust, and willingness to self-disclose. Journal of Multicultural Counseling and Development, 19, 65-73.

Cassidy, D. A., Bove, C. M. (1998). Factors perceived to influence parental decision-making regarding presymptomatic testing of children at risk for treatable adult-onset conditions. Issues in Comprehensive Pediatric Nursing, 21, 19-34.

Cavanagh, M. E. (1990). The counseling experience: A theoretical and practical approach. Prospect Heights, IL: Waveland Press.

Cerney, M. S. (1985). Countertransference revisited. Journal of Counseling and Development, 63, 362-364.

Chapple, A., Campion, P., May, C. (1997). Clinical terminology: Anxiety and confusion amongst families undergoing genetic counseling. Patient Education and Counseling, 32, 81-91.

Cheston, S. E. (1991). Making effective referrals. New York: Gardner Press.

Clark, A. J. (1991). The identification and modification of defense mechanisms in counseling. Journal of Counseling and Development, 69, 231-236.

Corey, G. (1996). Theory and practice of counseling and psychotherapy, 5th edition, Pacific Grove, CA: Brooks/Cole.

Corey, G., Schneider-Corey, M., Callanan, P. (1984). Issues and ethics in the helping professions. Monterey, CA: Brooks/Cole.

Cormier, L. S., Hackney, H. (1987). The professional counselor: A process guide to helping. Englewood Cliffs, NJ: Prentice-Hall.

Cormier, W. H., Cormier, L. S. (1991). Interviewing strategies for helpers: fundamental skills and cognitive behavioral interventions (3rd ed.). Pacific Grove, CA: Brooks/Cole.

Crisham, P. (1985). How can I do what's right? Nursing Management, 16(3), 42A-G.

Danish, S. J., D'Augelli, A. R. (1983). Helping skills II: life development intervention. New York: Human Sciences Press.

Danish, S. J., D'Augelli, A. R., Hauer, A. L. (1980). Helping skills: a basic training program (2nd ed.). New York: Human Sciences Press.

Danish, S. J., Hauer, A. L., D'Augelli, A. R. (1978). Helping skills: A basic training program. New York: Human Sciences Press.

Davis, B. B., Wood, L., Wilson, R. (1983). ABC's of teaching with excellence: a Berkeley compendium of suggestions for teaching with excellence. Berkeley: Office of Educational Development, University of California, Berkeley.

Davis, B. J., Voegtle, K. H. (1994). Culturally competent health care for adolescents. Chicago: Department of Adolescent Health, American Medical Association, 19-24.

DeCapua, A., Findlay Dunham, J. (1993). Strategies in the discourse of advice. Journal of Pragmatics, 20, 519-531.

Diller, L. H. (1986). On giving good advice successfully. Family Systems Medicine, 4, 78-90.

Dinklage, L. B. (1966). Adolescent choice and decision-making: a review of decision-making models and issues in relation to some developmental stage tasks of adolescence. Cambridge: Harvard University.

Djurdjinovic, L. (1998). Psychosocial counseling. In: Baker, D. L., Schuette, J. L., Uhlmann, W. R., eds. A guide to genetic counseling. New York: John Wiley & Sons.

Dougherty, A. M., Henderson, B. B., Lindsey, B. (1997). The effectiveness of direct versus indirect confrontation as a function of stage of consultation: results of an exploratory investigation. Journal of Educational and Psychological Consultation, 8, 361-372.

Downing, C. J. (1985). Referrals that work. The School Counselor, 32, 242-246.

Duan, C., Hill, C. E. (1996). Theoretical confusions in the construct of empathy: A review of the literature. Journal of Counseling Psychology, 43, 261-274.

Egan, G. (1994). The skilled helper (5th ed.). Monterey, CA: Brooks/Cole.

Ehrlich, R. P., D'Augelli, A. R., Danish, S. J. (1979). Comparative effectiveness of six counselor verbal responses. Journal of Counseling Psychology, 26, 390-398.

Ekman, P. (1972). Universal and cultural differences in facial expressions of emotions. In: Cole, J., ed. Nebraska symposium on motivation, vol 19. Lincoln, NE: University of Nebraska Press, 207-283.

Ekman, P., Friesen, W. V. (1969). Non-verbal leakage and clues to deception. Psychiatry, 32, 88-106.

Ekman, P., Friesen, W. V. (1984). Unmasking the face (reprint ed.). Palo Alto, CA: Consulting Psychologists Press.

Elliott, R. (1985). Helpful and nonhelpful events in brief counseling interviews: an empirical taxonomy. Journal of Counseling Psychology, 32, 301-322.

Erlanger, M. A. (1990). Using the genogram with the older client. Journal of Mental Health Counseling, 12, 321-331.

Eunpu, D. L. (1997). Systemically based psychotherapeutic techniques in genetic counseling. Journal of Genetic Counseling, 6, 1-20.

Faulkner, A., Argent, F., Jones, A., O'Keefe, C. (1995). Improving the skills of doctors in giving distressing information. Medical Education, 29, 303-307.

Fine, B. A. (1993). The evolution of nondirectiveness in genetic counseling and implications of the Human Genome Project. In: Bartels, D. M., LeRoy, B. S., Caplan, A., eds. Prescribing our future: ethical challenges in genetic counseling. New York: Aldine deGruyter, 101-117.

Fine, S. F., Glasser, P. H. (1996). First interview: establishing an effective helping relationship. Thousand Oaks, CA: Sage.

Fisher, N. L. (1996). Cultural and ethnic diversity: a guide for genetics professionals. Baltimore, MD: Johns Hopkins University Press.

Fisher, N. L., Lew, L. (1996). Culture of the countries of Southeast Asia. In: Fisher, N. L., ed. Cultural and ethnic diversity: a guide for genetics professionals. Baltimore, MD: Johns Hopkins University Press, 113-128.

Flash, P., Tzenis, C., Waller, A. (1995). Helpfulness of feedback given you about your performance. In: Using student evaluations to increase classroom effectiveness. Minneapolis, MN: Faculty and Teaching Assistant Enrichment Program, University of Minnesota, 58-61.

Fontaine, J. H., Hammond, N. L. (1994). Twenty counseling maxims. Journal of Counseling and Development, 73, 223-226.

Frets, P. G., Duivenvoorden, H. J., Verhage, F., Niermeijer, M. F. (1992). The reproductive decision-making process after genetic counseling: psychosocial aspects. Birth Defects: Original Article Series, 28, 21-28.

Gartner, J., Harmatz, M., Hohmann, A., Larson, D., Fishman Gartner, A. (1990). The effect of client and counselor values on clinical judgment. Counseling and Values, 35, 58-62.

Geldard, D. (1989). Basic personal counseling: a training manual for counselors. Springfield, IL: Charles C. Thomas.

Geller, G., Holtzman, M. A. (1991). Implications of the Human Genome Initiative for the primary care physician. Bioethics, 5(4), 319-325.

Gintner, G. (1996, May). Handling anger before it handles you. Counseling Today, 15-16.

Gladstein, G. (1983). Understanding empathy: integrating counseling, development and social psychology perspectives. Journal of Counseling Psychology, 30, 467-482.

Goldsmith, D. J., Fitch, K. (1997). The normative context of advice as social support. Human Communication Research, 23, 454-476.

Goodyear, R. K., Bernard, J. M. (1998). Clinical supervision: lessons from the literature. Counselor Education and Supervision, 38, 6-22.

Gravely Moss, C. E. (1985). Utilizing effective communication skills in crisis intervention. Emotional First Aid, 2, 9-16.

Gray, C. A., McCarthy Veach, P., Jones, K. R., Goreczny, A., Hoss, M. (2000, May). Addressing genetic issues: The interface of psychotherapy and genetic counseling. Minnesota Psychologist, 8-10.

Green, J. M., Murton, F. E. (1993). Duchenne muscular dystrophy: the experiences of 158 families. Report published by the Centre for Family Research, University of Cambridge.

Green, R. M., Thomas, A. M. (1997). Whose gene is it? A case discussion about familial conflict over genetic testing for breast cancer. Journal of Genetic Counseling, 6, 245-254.

Greeson, C., LeRoy, B. S., McCarthy Veach, P. (2001). A qualitative investigation of Somali immigrant perceptions of disability: implications for traditional genetic counseling practice. Journal of Genetic Counseling, 10, 359-378.

Hackney, H. L., Cormier, L. S. (1996). The professional counselor: A process guide to helping (3rd ed.). Needham Heights, MA: Prentice-Hall.

Hall, J. A., Roter, D. L., Katz, N. R. (1988). Meta-analysis of correlates of provider behavior in medical encounters. Medical Care, 26, 657-675.

Hallowell, N., Statham, H., Murton, F., Green, J., Richards, M. (1997). Talking about chance: the presentation of risk information during genetic counseling for breast and ovarian cancer. Journal of Genetic Counseling, 6, 269-286.

Handelman, L., Menahem, S., Eisenbruch, M. (1989). Transcultural understanding of a hereditary disorder: mucopolysaccharidosis VI in a Vietnamese family. Clinical Pediatrics, 28, 470-473.

Hanna, F. J., Hanna, C. A., Keys, S. G. (1999). Fifty strategies for counseling defiant, aggressive adolescents: reaching, accepting, and relating. Journal of Counseling and Development, 77, 395-404.

Harper, P. S. (1993). Insurance and genetic testing. Lancet, 341, 224-227.

Harper, R. G., Wiens, A. N., Matarazzo, J. D. (1978). Nonverbal communication: the state of the art. New York: John Wiley & Sons.

Harris, G. A., Watkins, D. (1987). Counseling the involuntary and resistant client. College Park, MD: American Correctional Association.

Hendrick, G. (1977). What do I do after they tell me how they feel? Personnel and Guidance Journal, 55, 249-252.

Hendrickson, S. M., McCarthy Veach, P., LeRoy, B. S. (2002). A qualitative investigation of student and supervisor perceptions of live supervision in genetic counseling. Journal of Genetic Counseling, 11, 25-50.

Hill, C. E., Helms, J. E., Tichenor, V., Spiegel, S. B., O'Grady, K. E., Perry, E. S. (1988). The effects of therapist response modes in brief psychotherapy. Journal of Counseling Psychology, 35, 222-233.

Hill, C. E., O'Brien, K. M. (1999). Helping skills: Facilitating exploration, insight, and action. Washington, DC: American Psychological Association.

Hjelle, D. J., Ziegler, L. A. (1984). Personality theories (2nd ed.). New York, McGraw-Hill.

Hopkins, D. (1997, February). Scarves, swamps and spiral staircases: Using metaphor in grief counseling. Counseling Today, 24.

Hughes, J. N., Erchul, W. P., Yoon, J., Jackson, T., Henington, C. (1997). Consultant use of questions and its relationship to consultee evaluation of effectiveness. Journal of School Psychology, 35, 281-297.

Hultman, K. E. (1976). Values as defenses. Personnel and Guidance Journal, 54, 269-271.

Huys, J., Evers-Kiebooms, G., d'Ydewalle, G. (1992). Decision making in the context of genetic risk: the use of scenarios. Birth Defects: Original Article Series, 28, 17-20.

Ishiyama, F. I. (1995). Culturally dislocated clients: self-validation and cultural conflict issues and counselling implications. Canadian Journal of Counselling, 29, 262-275.

Israel, J., Cunningham, M., Thumann, H., Shaver Arnos, K. (1992). Genetic counseling for deaf adults: Communication /language and cultural considerations. Journal of Genetic Counseling, 1, 135-153.

Ivey, A. E. (1994). Intentional interviewing and counseling: facilitating client development in a multicultural society (3rd ed.). Pacific Grove, CA: Brooks/Cole.

Izard, C. E. (1977) Human emotions. New York: Plenum.

Jacobson, G., McCarthy Veach, P., LeRoy, B. S. (2001). A survey of genetic counselors' use of informed consent documents for prenatal genetic counseling. Journal of Genetic Counseling, 10, 3-24.

Jecker, N. S., Carrese, J. A., Pearlman, R. A. (1995). Caring for patients in cross-cultural settings. Hastings Center Report, 25, 6-14.

Johnson, D. W., Johnson, R. T., Smith, K. A. (1991). Active learning: cooperation in the college classroom. Edina, MN: Interaction.

Josephs, L. (1997). Balancing empathy and interpretation in the analytic process. Issues in Psychoanalytic Psychology, 19, 5-25.

Kennedy, A. L. (2000). Supervision for practicing genetic counselors: An overview of models. Journal of Genetic Counseling, 9, 391-398.

Kessler, S. (1979). Genetic counseling: psychological dimensions. New York: Academic Press, 17-33.

Kessler, S. (1980). The psychological paradigm shift in genetic counseling. Social Biology, 27, 167-185.

Kessler, S. (1992a). Psychological aspects of genetic counseling. VII. Thoughts on directiveness. Journal of Genetic Counseling, 1, 9-17.

Kessler, S. (1992b). Psychological aspects of genetic counseling. VIII. Suffering and countertransference. Journal of Genetic Counseling, 1, 303-308.

Kessler, S. (1997a). Psychological aspects of genetic counseling: IX. Teaching and counseling. Journal of Genetic Counseling, 6, 287-295.

Kessler, S. (1997b). Psychological aspects of genetic counseling: X. Advanced counseling techniques. Journal of Genetic Counseling, 6, 379-392.

Kessler, S. (1997c). Psychological aspects of genetic counseling: XI. Nondirectiveness revisited. American Journal of Medical Genetics, 72, 164-171.

Kessler, S. (1998). Psychological aspects of genetic counseling: XIII. More on counseling skills. Journal of Genetic Counseling, 7, 263-278.

Kessler, S. (1999). Psychological aspects of genetic counseling: XIII. Empathy and decency. Journal of Genetic Counseling, 8(6), 333-344.

Knox, S., Hess, S. A., Petersen, D. A., Hill, C. E. (1997). A qualitative analysis of client perceptions of the effects of helpful therapist self-disclosure in long-term therapy. Journal of Counseling Psychology, 44, 274-283.

Kopinsky, S. M. (1992). Letters to the editor: Value-based directiveness in genetic counseling. Journal of Genetic Counseling, 1, 345-348.

Kramer, S. A. (1990). Positive endings in psychotherapy. San Francisco, CA: Jossey-Bass.

Krueger, R. A. (1994). Focus groups: A practical guide for applied research (2nd ed.). Newbury Park, CA: Sage.

Lamb, D. H., Presser, N. R., Pfost, K. S., Baum, M. C., Jackson, V. R., Jarvis, P. A. (1987). Confronting professional impairment during the internship: identification, due process, and remediation. Professional Psychology: Research and Practice, 18, 597-603.

Laning, W., Carey, J. (1987). Systematic termination in counseling. Counselor Education and Supervision, 27, 168-173.

Larrabee, M. J. (1982). Working with reluctant clients through affirmation techniques. Personnel and Guidance Journal, 61, 105-109.

Leaman, D. R. (1978). Confrontation in counseling. Personnel and Guidance Journal, 56, 630-633.

Lewin, K. (1947). Frontiers in group dynamics. Human Relations, 1, 5-42.

Liburd, R. (1978). Defense mechanisms versus openness to experience: implications for counseling. Canadian Counsellor, 13, 33-36.

Lippman-Hand, A., Fraser, F. C. (1979a). Genetic counseling—the postcounseling period: I. Parents' perceptions of uncertainty. American Journal of Medical Genetics, 4, 51-71.

Lippman-Hand, A., Fraser, F. C. (1979b). Genetic counseling: Provision and reception of information. American Journal of Medical Genetics, 3, 113-127.

Lipson, J. G., Meleis, A. I. (1983). Issues in health care of Middle Eastern patients. Western Journal of Medicine, 139, 854-861.

Loughead, T. A. (1992). Freudian repression revisited: the power and pain of shame. International Journal for the Advancement of Counselling, 15, 127-136.

Macran, S., Shapiro, D. A. (1998). The role of personal therapy for therapists: a review. British Journal of Medical Psychology, 71, 13-25.

Macran, S., Stiles, W. B., Smith, J. A. (1999). How does personal therapy affect therapists' practice? Journal of Counseling Psychology, 46, 419-431.

Maley, J. A. (1994). An ethics casebook for genetic counselors. Wallingford, PA: NSGC.

Marks, J. H. (1993). The training of genetic counselors: origins of a psychosocial model. In: Bartels, D. M., LeRoy, B. S., Caplan, A. L., eds. Prescribing our future: Ethical challenges in genetic counseling. New York: Walter de Gruyter, 15-24.

Markus, H. R., Kitayama, S. (1994). The cultural shaping of emotion: a conceptual framework. In: Kitayama, S., Markus, H. R., eds. Emotion and culture: empirical studies of mutual influence. Washington, DC: American Psychological Association, 339-351.

Martin, D. G. (2000). Counseling and therapy skills (2nd ed.). Prospect Heights, IL: Waveland Press.

Marziali, E. (1988, January). The first session: an interpersonal encounter. Social Casework: Journal of Contemporary Social Work, 23-27.

Mayfield, W. A., Kardash, C. M., Kivlighan, D. M., Jr. (1999). Differences in experienced and novice counselors' knowledge structures about clients: implications for case conceptualization. Journal of Counseling Psychology, 46, 504-514.

McCarthy, P. R. (1979). Differential effects of self-disclosing versus self-involving counselor statements across counselor-client gender pairings. Journal of Counseling Psychology, 26, 538-541.

McCarthy, P. R. (1982). Differential effects of self-referent responses and counselor status. Journal of Counseling Psychology, 29, 125-131.

McCarthy, P. R., Danish, S. J., D'Augelli, A. R. (1977). A follow-up evaluation of helping skills training. Counselor Education and Supervision, 17, 29-35.

McCarthy, P., LeRoy, B. S. (1998). Student supervision. In: Baker, D., Schuette, J. L., Uhlmann, W. R., eds. A guide to genetic counseling. New York: John Wiley & Sons, 295-330.

McCarthy, P., Oakes, L. (1998). Blank screen or open book? A reminder about balancing self-disclosure in psychotherapy. Voices: The Art and Science of Psychotherapy, 34, 60-68.

McCarthy Veach, P., Bartels, D. M., LeRoy, B. S. (2001). Ethical and professional challenges posed by patients with genetic concerns: a report of focus group discussions with genetic counselors, physicians, and nurses. Journal of Genetic Counseling, 10, 97-119.

McCarthy Veach, P., Truesdell, S., LeRoy, B. S., Bartels, D. M. (1999). Client perceptions of the impact of genetic counseling: an exploratory study. Journal of Genetic Counseling, 8, 191-216.

McGoldrick, M., Gerson, R. (1985). Genograms in family assessment. New York: W. W. Norton.

McSee, G. S. (1985). "Hearing" nonverbal cues in controlling aggressive clients. Emotional First Aid, 2, 47-53.

Mealy, L. (1984). Decision making and adjustment in genetic counseling. Health and Social Work, 9, 124-133.

Mehrabian, A. (1976). Public places and private spaces. New York: Basic Books.

Michie, S., Weinman, J., Marteau, T. M. (1998). Genetic counselors' judgments of patient concerns: concordance and consequences. Journal of Genetic Counseling, 7, 219-231.

Murray, R. F., Jr. (1976). Psychosocial aspects of genetic counseling. Social Work in Health Care, 2, 13-23.

Murray, T. H. (1992). Genetics and the moral mission of health insurance. Hastings Center Report, 22, 12-17.

National Society of Genetic Counselors Web site (2001). http://www.nsgc.org/.

Nutt-Williams, E., Judge, A. B., Hill, C. E., Hoffman, M. A. (1997). Experiences of novice therapists in prepracticum: trainees', clients', and supervisors' perceptions of therapists' personal reactions and management strategies. Journal of Counseling Psychology, 44, 390-399.

Olsen, D. H., Claiborn, C. D. (1990). Interpretation and arousal in the counseling process. Journal of Counseling Psychology, 37, 131-137.

Ormerod, J. J., Huebner, S. E. (1988). Crisis intervention: Facilitating parental acceptance of a child's handicap. Psychology in the Schools, 25, 422-428.

Ostergren, J. (1991). Relationships among English performance, self-efficacy, anxiety, and depression for Hmong refugees. Unpublished doctoral dissertation. University of Minnesota, Minneapolis, MN.

Palmer, C. G. S., Sainfort, F. (1993). Toward a new conceptualization and operationalization of risk perception within the genetic counseling domain. Journal of Genetic Counseling, 2, 275-294.

Papadopoulos, L., Bor, R., Stanion, P. (1997). Genograms in counselling practice: a review (part 1). Counselling Psychology Quarterly, 10, 17-28.

Pedersen, P. B. (1991). Multiculturalism as a generic approach to counseling. Journal of Counseling and Development, 70, 6-12.

Pedersen, P. B., Ivey, A. (1993). Culture-centered counseling and interviewing skills: a practical guide. Westport, CN: Praeger.

Pedersen, P. (1995). The culture-bound counsellor as an unintentional racist. Canadian Journal of Counselling, 29, 197-205.

Peebles, M. J. (1980). Personal therapy and ability to display empathy, warmth and genuineness in psychotherapy. Psychotherapy: Theory, Research and Practice, 17, 258-262.

Pinkerton, R. S., Rockwell, W. K. (1990). Termination in brief psychotherapy: the case for an eclectic approach. Psychotherapy, 27, 362-365.

Punales-Morejon, D., Penchaszadeh, V. B. (1992). Psychosocial aspects of genetic counseling: cross-cultural issues. Birth Defects: Original Article Series, 28, 11-15.

Quintana, S. M. (1993). Toward an expanded and updated conceptualization of termination: implications of short-term, individual psychotherapy. Professional Psychology: Research and Practice, 24, 426-432.

Reed, S. C. (1955). Counseling in medical genetics. Philadelphia, PA: WB Sanders.

Reed, S. C. (1980). Counseling in medical genetics, 3rd ed. New York: Alan R. Liss.

Rich, D. E. (1999). When your client's baby dies. Journal of Couples Therapy, 8, 49-60.

Ridley, C. R. (1995). Overcoming unintentional racism in counseling and therapy: a practitioner's guide to intentional intervention. Thousand Oaks, CA: Sage.

Ridley, C. R., Lingle, D. W. (1996). Cultural empathy in multicultural counseling. In: Pedersen, P. B., Draguns, J. G., Lonner, W. J., Trimble, L. D., eds. Counseling across cultures (4th ed.). Thousand Oaks, CA: Sage, 21-46.

Rogers, C. R. (1992). The necessary and sufficient conditions of therapeutic personality change. Journal of Consulting and Clinical Psychology, 60, 827-832.

Rogers, J., Durkin, M. (1984). The semi-structured genogram interview: I. Protocol, II. Evaluation. Family Systems Medicine, 2, 176-187.

Rose, R., Humm, E., Hey, K., Jones, L., Huson, S. M. (1999). Family history taking and genetic counseling. Family Practice, 16, 78-83.

Sagi, M., Kaduri, L., Zlotogora, J., Petetz, T. (1998). The effect of genetic counseling on knowledge and perceptions regarding risks for breast cancer. Journal of Genetic Counseling, 7, 417-434.

Salzman, M. (1995). Attributional discrepancies and bias in cross-cultural interactions. Journal of Multicultural Counseling and Development, 23, 181-193.

Sanders, N. M. (1966). Classroom questions: what kind? New York: Harper and Row.

Schuette, J. L., Bennett, R. L. (1998). Lessons in history: obtaining the family history and constructing a pedigree. In: Baker, D. L., Schuette, J. L., Uhlmann, W. R., eds. A guide to genetic counseling. New York: John Wiley & Sons, 27-54.

Segall, M. H. (1986). Culture and behavior: psychology in global perspective. Annual Review of Psychology, 37, 523-564.

Selman, R. L. (1980). The growth of interpersonal understanding. New York: Academic Press.

Silver, E. (1991). Should I give advice? A systemic view. Journal of Family Therapy, 13, 295-309.

Simone, D. H., McCarthy, P., Skay, C. L. (1998). An investigation of client and counselor variables that influence likelihood of counselor self-disclosure. Journal of Counseling and Development, 76, 174-182.

Simonoff, E. (1998). Genetic counseling in autism and pervasive developmental disorders. Journal of Autism and Developmental Disorders, 28, 447-456.

Skouholt, T. S. (2000). The resilient practitioner: Burnout prevention and self-care. Boston: Allyn and Bacon.

Sorenson, J. R. (1976). From social movements to clinical medicine: the role of law and the medical profession in regulating applied human genetics. In: Genetics and the law. New York: Plenum Press.

Sorenson, J. R. (1993). Genetic counseling: values that have mattered. In: Bartels, D. M., LeRoy, B. S., Caplan, A. L., eds. Prescribing our future: ethical challenges in genetic counseling. New York: Walter deGruyter, 3-14.

Stanion P., Papadopoulos, L., Bor, R. (1997). Genograms in counselling practice: constructing a genogram (part 2). Counselling Psychology Quarterly, 10, 139-148.

Sue, D. W., Sue, D. (1990). Counseling the culturally different theory and practice (2nd ed.). New York: John Wiley & Sons.

Sue, S., Zane, N. (1987). The role of culture and cultural techniques in psychotherapy: a critique and reformulation. American Psychologist, 42, 37-45.

Turock, A. (1980). Immediacy in counseling: recognizing clients' unspoken messages. Personnel and Guidance Journal, 59, 168-172.

Van Bezooijen, R., Otto, S. A., Heenan, T. A. (1983). Recognition of vocal expressions of emotion: a three-nation study to identify universal characteristics. Journal of Cross-Cultural Psychology, 14, 387-406.

Van Spijker, H. G. (1992). Support in decision making processes in the post-counseling period. Birth Defects: Original Article Series, 28, 29-35.

Volz, J. (2000). Clinician, heal thyself. American Psychological Association Monitor, 31, 46-47.

Vontress, C. E. (1988). Social class influences on counseling. In: Hayes, R., Aubrey, R., eds. New directions for counseling and human development. Denver, CO: Love.

Vriend, J., Kottler, J. A. (1980). Initial interview checklist increases counselor effectiveness. Canadian Counsellor, 14, 153-155.

Walker, A. P. (1998). The practice of genetic counseling. In: Baker, D. L., Schuette, J. L., Uhlmann, W. R., eds. A guide to genetic counseling. New York: John Wiley & Sons, 1-26.

Wang, V. (1993). Handbook of cross-cultural genetic counseling. Unpublished manual.

National Society of Genetic Counselors Code of Ethics

Preamble

Genetic counselors are health professionals with specialized education, training, and experience in medical genetics and counseling. The National Society of Genetic Counselors (NSCC) is an organization that furthers the professional interests of genetic counselors, promotes a network for communication within the profession, and deals with issues relevant to human genetics.

With the establishment of this code of ethics the NSGC affirms the ethical responsibilities of its members and provides them with guidance in their relationships with self, clients, colleagues, and society. NSGC members are expected to be aware of the ethical implications of their professional actions and to adhere to the guidelines and principles set forth in this code.

Introduction

A code of ethics is a document which attempts to clarify and guide the conduct of a professional so that the goals and values of the profession might best be served. The NSGC Code of Ethics is based upon relationships. The relationships outlined in this code describe who genetic counselors are for themselves, their clients, their colleagues, and society. Each major section of this code begins with an explanation of one of these relationships, along with some of its values and characteristics. Although certain values are found in more than one relationship, these common values result in different guidelines within each relationship.

No set of guidelines can provide all the assistance needed in every situation, especially when different relationships appear to conflict. Therefore, when considered appropriate for this code, specific guidelines for prioritizing the relationships have been stated. In other areas, some ambiguity remains, allowing for the experience of genetic counselors to provide the proper balance in responding to difficult situations.

Section I: Genetic Counselors Themselves

Genetic counselors value competence, integrity, dignity, and self-respect in themselves as well as in each other. Therefore, in order to be the best possible human resource to themselves, their clients, their colleagues, and society, genetic counselors strive to

1. Seek out and acquire all relevant information required for any given situation.
2. Continue their education and training.
3. Keep abreast of current standards of practice.
4. Recognize the limits of their own knowledge, expertise, and therefore competence in any give situation.
5. Be responsible for their own physical and emotional health as it impacts on tier professional performance.

Section II: Genetic Counselors and Their Clients

The counselor-client relationship is based on values of care and respect for the client's autonomy, individuality, welfare, and freedom. The primary concern of genetic counselors is the interests of their clients. Therefore, genetic counselors strive to

1. Equally serve all who seek services.
2. Respect their client's beliefs, cultural traditions, inclinations, circumstances, and feelings.
3. Enable their clients to make informed independent decisions, free of coercion, by providing or illuminating the necessary facts and clarifying the alternatives and anticipated consequences.
4. Refer clients to other competent professionals when they are unable to support the clients.
5. Maintain as confidential any information received from clients, unless released by the client.
6. Avoid the exploitation of their clients for personal advantage, profit, or interest.

Section III: Genetic Counselors and Their Colleagues

The genetic counselors' relationships with other genetic counselors, genetic counseling students, and health professionals are based on mutual respect, caring, cooperation, support, and a shared loyalty to their professions and goals. Therefore, genetic counselors strive to

1. Foster and protect their relationship with other genetic counselors and genetic counseling students by establishing mechanisms for peer support.
2. Encourage ethical behavior of colleagues.
3. Recognize the traditions, practices, and areas of competence of other health professionals and cooperate with them in providing the highest quality of service.
4. Work with their colleagues to research consensus when issues arise about the role responsibilities of various team members so that clients receive the best possible services.

Section IV: Genetic Counselors and Society

The relationships of genetic counselors to society include interest and participation in activities that have the purpose of promoting the well-being of society. Therefore, genetic counselors strive to

1. Keep abreast of societal developments that may endanger the physical and psychological health of individuals.
2. Participate in activities necessary to bring about socially responsible change.
3. Serve as a source of reliable information and expert opinion for policy makers and public officials.
4. Keep the public informed and educated about the impact on society of new technological and scientific advances and the possible changes in society that may result from the application of these findings.
5. Prevent discrimination on the basis of race, sex, sexual orientation, age, religion, genetic status, or socioeconomic status.
6. Adhere to laws and regulations of society. However, when such laws are in conflict with the principles of the profession, genetic counselors work toward change that will benefit the public interest.

Acknowledgment

The above appendix was reprinted with permission of the National Society of Genetic Counselors (NSGC) June, 1991, Human Sciences Press.

American Board of Genetic Counseling Practice-Based Competencies

Introduction

An entry-level genetic counselor must demonstrate the practice-based competencies listed below to manage a genetic counseling case before, during, and after the clinic visit or session. Therefore, the didactic and clinical training components of a curriculum must support the development of competencies that are categorized into the following domains: Communication Skills; Critical-Thinking Skills; Interpersonal, Counseling, and Psychosocial Assessment Skills; and Professional Ethics and Values. Some competencies may pertain to more than one domain. These domains represent practice areas that define activities of a genetic counselor. The italicized facet below each competency elaborates on skills necessary for achievement of each competency. These elaborations should assist program faculty in curriculum planning, development, and program and student evaluation.

Domain I: Communication Skills

a. Can establish a mutually agreed upon genetic counseling agenda with the client.

The student is able to contract with a client or family throughout the relationship; explain the genetic counseling process; elicit expectations, perceptions and knowledge; and establish rapport through verbal and nonverbal interaction.

b. Can elicit an appropriate and inclusive family history.

The student is able to construct a complete pedigree; demonstrate proficiency in the use of pedigree symbols, standard notation, and nomenclature; structure questioning for the individual case and probable diagnosis; use interviewing skills; facilitate recall for symptoms and pertinent history by pursuing a relevant path of inquiry; and in the course of this interaction, identify family dynamics, emotional responses, and other relevant information.

c. Can elicit pertinent medical information including pregnancy, developmental, and medical histories.

The student is able to apply knowledge of the inheritance patterns, etiology, clinical features, and natural history of a variety of genetic disorders, birth defects, and other conditions; obtain appropriate medical histories; identify essential medical records and secure releases of medical information.

d. Can elicit a social and psychosocial history.

The student is able to conduct a client or family interview that demonstrates an appreciation of family systems theory and dynamics. The student is able to listen effectively, identify potential strengths and weaknesses, and assess individual and family support systems and coping mechanisms.

e. Can convey genetic, medical, and technical information including, but not limited to, diagnosis, etiology, natural history, prognosis, and treatment/management of genetic conditions and/or birth defects to clients with a variety of educational, socioeconomic, and ethnocultural backgrounds.

The student is able to demonstrate knowledge of clinical genetics and relevant medical topics by effectively communicating this information in a given session.

f. Can explain the technical and medical aspects of diagnostic and screening methods and reproductive options including associated risks, benefits, and limitations.

The student is able to demonstrate knowledge of diagnostic and screening procedures and clearly communicate relevant information to clients. The student is able to facilitate the informed-consent process. The student is able to determine client comprehension and adjust counseling accordingly.

g. Can understand, listen, communicate, and manage a genetic counseling case in a culturally responsive manner.

The student can care for clients using cultural self-awareness and familiarity with a variety of ethnocultural issues, traditions, health beliefs, attitudes, lifestyles, and values.

h. Can document and present case information clearly and concisely, both orally and in writing, as appropriate to the audience.

The student can present succinct and precise case-summary information to colleagues and other professionals. The student can to write at an appropriate level for clients and professionals and produce written documentation within a reasonable timeframe. The student can demonstrate respect for privacy and confidentiality of medical information.

i. Can plan, organize, and conduct public and professional education programs on human genetics, patient care, and genetic counseling issues.

The student is able to identify educational needs and design programs for specific audiences, demonstrate public speaking skills, use visual aids, and identify and access supplemental educational materials.

Domain II: Critical-Thinking Skills

a. Can assess and calculate genetic and teratogenic risks.

The student is able to calculate risks based on pedigree analysis and knowledge of inheritance patterns, genetic epidemiologic data, and quantitative genetics principles.

b. Can evaluate a social and psychosocial history.

The student demonstrates understanding of family and interpersonal dynamics and can recognize the impact of emotions on cognition and retention, us well as the need for intervention and referral.

c. Can identify, synthesize, organize, and summarize pertinent medical and genetic information for use in genetic counseling.

The student is able to use a variety of sources of information including client/family member(s), laboratory results, medical records, medical and genetic literature and computerized databases. The student is able to analyze and interpret information that provides the basis for differential diagnosis, risk assessment and genetic testing. The student is able to apply knowledge of the natural history and characteristics/symptoms of common genetic conditions.

d. Can demonstrate successful case management skills.

The student is able to analyze and interpret medical, genetic and family data; to design, conduct, and periodically assess the case management plan; arrange for testing; and follow up with the client, laboratory, and other professionals. The student should demonstrate understanding of legal and ethical issues related to privacy and confidentiality in communications about clients.

e. Can assess client understanding and response to information and its implications to modify a counseling session as needed.

The student is able to respond to verbal and nonverbal cues and to structure and modify information presented to maximize comprehension by clients.

f. Can identify and access local, regional, and national resources and services. The student is familiar with local, regional, and national support groups and other resources, and can access and make referrals to other professionals and agencies.

The student is familiar with local, regional, and national support groups and other resources, and can access and make referrals to other professionals and agencies.

g. Can identify and access information resources pertinent to clinical genetics and counseling.

The student is able to demonstrate familiarity with the genetic, medical and social-science literature, and on-line databases. The student is able to review the literature and synthesize the information for a case in a critical and meaningful way.

Domain III: Interpersonal, Counseling, and Psychosocial Assessment Skills

a. Can establish rapport, identify major concerns, and respond to emerging issues of a client or family.

The student is able to display empathic listening and interviewing skills, and address clients' concerns.

b. Can elicit and interpret individual and family experiences, behaviors, emotions, perceptions, and attitudes that clarify beliefs and values.

The student is able to assess and interpret verbal and nonverbal cues and use this information in the genetic counseling session. The student is able to engage clients in an exploration of their responses to risks and options.

c. Can use a range of interviewing techniques.

The student is able to identify, and select from a variety of communication approaches throughout a counseling session.

d. Can provide short-term, client-centered counseling and psychological support.

The student is able to assess clients' psychosocial needs and recognize psychopathology. The student can demonstrate knowledge of psychological defenses, family dynamics, family theory, crisis-intervention techniques, coping models, the grief process, and reactions to illness. The student can use open-ended questions; listen empathically, employ crisis-intervention skills; and provide anticipatory guidance.

e. Can promote client decisions-making in an unbiased, noncoercive manner.

The student understands the philosophy of nondirectiveness and is able to recognize his or her values and biases as they relate to genetic counseling issues. The student is able to recognize and respond to dynamics, such as countertransference, that may affect the counseling interaction.

f. Can establish and maintain inter- and intradisciplinary professional relationships as part of a health-care delivery team.

The student behaves professionally and understands the roles of other professionals with whom he or she interacts.

Domain IV: Professional Ethics and Values

a. Can act in accordance with the ethical, legal, and philosophical principles and values of the profession.

The student is able to recognize and respond to ethical and moral dilemmas arising in practice and seek assistance from experts in these areas. The student is able to identify factors that promote or hinder client autonomy. The student demonstrates an appreciation of the issues surrounding privacy, informed consent, confidentiality, real or potential discrimination, and other ethical/legal matters related to the exchange of genetic information.

  b. Can serve as an advocate for clients.

The student can understand clients' needs and perceptions and represent their interests in accessing services and responses from the medical and social service systems.

  c. Can introduce research options and issues to clients and families.

The student is able to critique and evaluate the risks, benefits, and limitations of client participation in research access information on new research studies; present this information clearly and completely to clients; and promote an informed-consent process.

  d. Can recognize his or her own limitations in knowledge and capabilities regarding medical, psychosocial, and ethnocultural issues and seek consultation or refer clients when needed.

The student demonstrates the ability to self-assess and to be self-critical. The student demonstrates the ability to respond to performance critique and integrates supervision feedback into his or her subsequent performance. The student is able to identify, and obtain appropriate consultative assistance for self and clients.

  e. Can demonstrate initiative for continued professional growth.

The student displays a knowledge of current standards of practice and shows independent knowledge-seeking behavior and lifelong learning.

From the American Board of Genetic Counseling. (1996). Journal of Genetic Counseling 5(3), 113-121.

Index